Magnetic Resonance in Food Science

From Molecules to Man

Magnetic Resonance in Food Science

From Molecules to Man

Edited by

I A Farhat
School of Biosciences, University of Nottingham, Loughborough, UK

P S Belton
School of Chemical Sciences and Pharmacy, University of East Anglia, Norwich, UK

G A Webb
c/o Burlington House, Piccadilly, London, UK

RSC Publishing

The proceedings of the 8th International Conference on the Applications of Magnetic Resonance in Food Science held at the University of Nottingham on 16-19 July 2006.

Special Publication No. 310

ISBN: 978-0-85404-340-8

A catalogue record for this book is available from the British Library

Published by The Royal Society of Chemistry,
Thomas Graham House, Science Park, Milton Road,
Cambridge CB4 0WF, UK

Registered Charity Number 207890

For further information see our web site at www.rsc.org

Printed by Henry Ling Ltd, Dorchester, Dorset. UK

Preface

The modern challenges of food science and technology require a better understanding of the utilisation of food by the human body and the determinants of food quality. We need to know how raw materials, processing, and changes during storage impact upon the sensory and nutritional attributes of foods. Magnetic resonance is well positioned to help address these challenges. This is being achieved through a combination of the following developments:

- Advances in hardware, in particular in terms of technically and economically achievable field strength.
- Advances in data analysis e.g. through the use of complex multivariate protocols.
- Affordability and robustness of instruments, widening the access to magnetic resonance.

The 8th *International Conference on the Applications of Magnetic Resonance in Food Science* was held at The University of Nottingham on the 16th - 19th July 2006. As the principal conference in the field of magnetic resonance in food sciences, the event attracts contributions from internationally acknowledged experts from academia and industry and an audience from all over the world. This edition was inaugurated by a lecture by Sir Peter Mansfield, Medicine Nobel prize laureate for his discoveries in magnetic resonance imaging.
The 8th edition of the proceedings is entitled *From Molecules to Man* in order to truly reflect the breath of the applications of nuclear magnetic resonance in all aspects of food sciences: from the studies of subtle molecular motion modes to macroscopic scale mass transfer during processing and from structure elucidation to authentication, quality and metabonomics.

This book is based on contributions to the conference technical programme. It is structured into four major sections:

- Food in the human body which includes MRI and metabonomics studies.
- Food quality which includes papers on animal metabonomics, structure of food systems, food stability and authentication.
- Food processing with particular emphasis on dynamic processes such as water migration and phase transformations.
- New techniques, novel data analysis and exploitation covering innovations in NMR methodologies, hardware and data analysis, for example using multivariate approaches.

We would like to thank the contributors to the book and the conference, the staff in the Division of Food Sciences and the Sir Peter Mansfield Magnetic Resonance Centre at the University of Nottingham who supported the event, in particular Mrs Val Street, and last but not least, The Royal Society of Chemistry for the realisation of the book.

The editors:

Imad Farhat
Firmenich, Geneva, Switzerland

Peter Belton
School of Chemical Sciences and Pharmacy, University of East Anglia, Norwich, UK

Graham Webb
c/o Burlington House, Piccadilly, London, UK

1992: Surrey, UK

1994: Aveiro, Portugal

1996: Nantes, France

1998: Norwich, UK

2000: Aveiro, Portugal

2002: Paris, France

2004: Copenhagen, Denmark

2006: Nottingham, UK

Contents

Food Processing

New Techniques and Novel Data Analysis and Exploitation

NMR IN FOODS: THE INDUSTRIAL PERSPECTIVE

J.P.M. van Duynhoven, A. Haiduc, F. van Dorsten and E. van Velzen

Unilever Food and Health Research Institute, Olivier van Noortlaan 120, 3130AC, Vlaardingen, The Netherlands. E-mail: john-van.duynhoven@unilever.com

1 INTRODUCTION

1.1 Role of NMR in the Foods Industry

Current State of the Industry

A main challenge for the foods industry for the next decades is to address the growing concerns of society with respect to public health. In the whole western world obesity, cardiovascular disease, arthrosis, and diabetes are on the rise (1,2) and the link with eating patterns is scientifically well established. The industry is now addressing the growing demand of consumers and legislators for products that promote sustainable wellbeing and health. In one route, the industry is responding by reducing high levels of fat, sugar and salt, since these are clearly linked to adverse health effects. Since, these ingredient critically determine taste and texture of the current generation of food products, the industry is now posing the challenge of re-designing their microstructures (3). In a second route, the industry is searching for food ingredients that actively promote health. Both routes require significant scientific and technological investments, in particular since new legislation is requiring that claims should be sustained by sound scientific evidence.

What to Measure

In order to be successful, the industry recognizes the key role of science and technology. Since 'all good science is measurement' (Helmholtz), this offers ample opportunities for NMR. Within the realm of measurement technologies, NMR takes a unique position due to its wide range of application areas. NMR can be deployed in diverse areas as the measurement of product meso- and microstructures, product compositions as well as the physiological and psychological impact of products on consumers.

How to Measure

Within science we are witnessing a shift from hypothesis testing towards hypothesis generation. This is particularly apparent within the emerging 'omics' life sciences (4,5). Such approaches can only be successful when measurement technologies are available that are not particularly biased by prior expectations. NMR has established itself a position in this field to its comprehensiveness and unbiased nature. In this review examples will be presented from the nutritional metabonomics, and we will also demonstrate how NMR can be applied in an analogous manner in the microstructural area.

Although NMR is still considered as a technique that requires expensive instrumentation, trained operators and data interpreters, non-NMR experts can already obtain useful data from benchtop NMR instruments. Developments in this area are speeding up, and we can now even envisage NMR (self)-measurements by consumers.

1.2 Define, Discover, Design, Deploy, Deliver

The competitive position of foods industries strongly relies on the efficiency with which they can bring new and healthy product innovations to the market. Within the innovation process one can roughly discern 5 phases: Define, Discover, Design, Deploy and Deliver (Figure 1A). In the Define phase, understanding of consumer needs is translated into outlines for novel products and services that eventually can be brought to market. In the Discovery phase, the food scientist acquires new insights that allow the design of novel food products. In the Design phase, ideas that were conceived in the Discovery phase are transformed into product prototypes. Subsequently, in the Deployment phase, food engineers scale up processes up to mass production. Another critical success factor is the efficiency of the supply chain, which Delivers food products to the consumer. This involves safe manufacturing and efficient distribution through a range of logistic channels (retail, food services etc.).

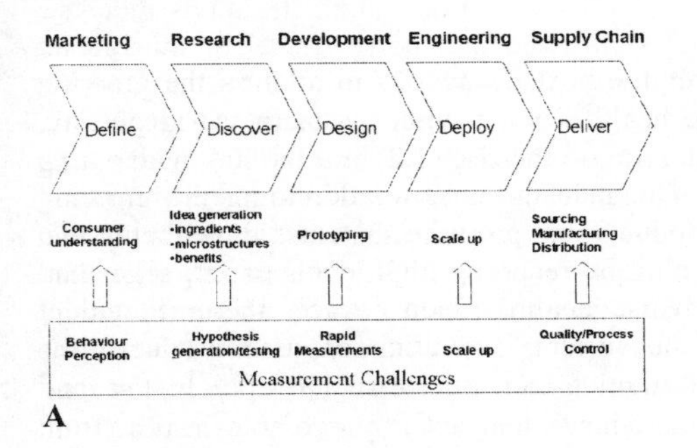

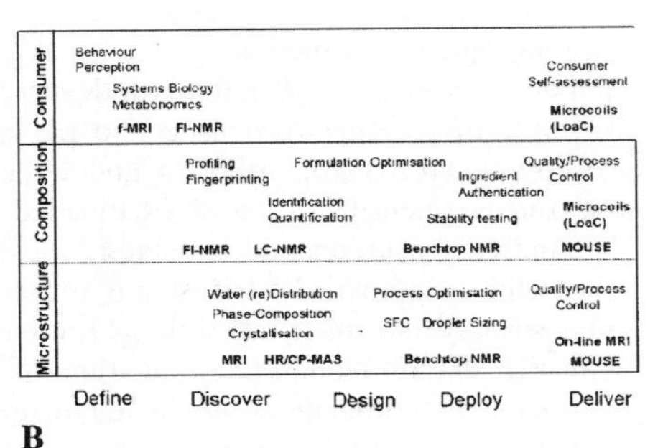

Figure 1 *(A) Schematic depiction of the industrial innovation process. (B) Overview of applications of NMR throughout the 5 innovation phases (horizontal) and main NMR/MRI application areas (vertical). Methods are indicated in bold, applications in italics (see text).*

Within the foods industry, NMR plays a unique role since it can make important contributions in all DDDDD phases. It can do so by providing insight in product composition and microstructure, and their interactions with the consumer. An overview of the vast and broad amount of NMR applications in the foods industry is given in Figure 1B. In this review, the industrial opportunities of NMR will be illustrated by two examples from the author's own practice. Two product concepts will be tracked through different innovation phases, involving measurements on both product and consumer level. Trends in the industrial field will be illustrated by recent work and reviews, but no attempt has been made to give an exhaustive overview.

2 HEALTH IMPACT OF FLAVONOIDS

2.1 Flavonoids

In their quest for 'natural' functional ingredients with beneficial health effects, the foods industries have developed a strong interest in polyphenols. These compounds have been associated with prevention of diseases, in particular cardiovascular disorders (6,7). Among the polyphenols, beneficial effects are best articulated for the flavonoids. Most evidence, however, is based on a limited number of biomarkers and clinical end-points. Flavonoids do have an impact on oxidative stress markers, but there is a growing awareness that the 'antioxidant theory' is a naïve simplification (6). Hence there is a need to pursue further research in this area.

2.2 Discovery: health benefits of flavonoids

Until recently, nutritional research was hypothesis-driven and departed from pre-identified markers and benefits. The relatively slow progress made in the last decades is driving a movement towards more exploratory, hypothesis generating 'omics' approaches, which assess living organisms in a holistic manner (8). Sofar most nutritional applications of metabonomics relied on NMR for unbiased profiling of body fluids (9,10), and used multivariate data analysis techniques (11) to recognise patterns and establish relations with accepted biomarkers. Metabonomics has taken a firm position within the foods industry since can provide direct feedback on health status and metabolic effects of nutritional interventions (12-15). Figure 2 shows examples of the type of information that can be obtained from nutritional intervention studies. In a (double blind) cross-over trial, volunteers were taking grape/wine extract and a placebo. The metabolic impact was assessed by measuring ^{1}H NMR spectra of body fluids collected from the volunteers. Figure 2A and 2B show that a grape/wine extract had a significant impact on the metabolite composition of these urine and plasma, respectively. Several metabolites that are responsible for the clustering in Figure 2 have been identified and could be attributed to both exogenous (xenobiotic) and endogenous effects. The exogenous impact mainly involved low molecular weight flavonoid degradation products. This points towards bioconversion by gut microflora, and is in line with metabonomic studies on the impact of other flavonoid sources (16-19). It has been recognised that bioavailability of intact dietary flavonoids is limited, and that the explanation of their beneficial effects may lie in secondary metabolites produced by gut microflora (6). The role of gut microflora in human health is gaining considerable interest, and metabonomics will be critical for gaining further insights (15,20,21). In the next years, when the hypotheses in this area will become more focussed, more sensitive and targeted metabolic profiling techniques based on mass spectrometry will be used ((22). Meanwhile, NMR will remain the preferred technique to obtain comprehensive and unbiased metabolic profiles, with minimal sample pre-treatment (23).

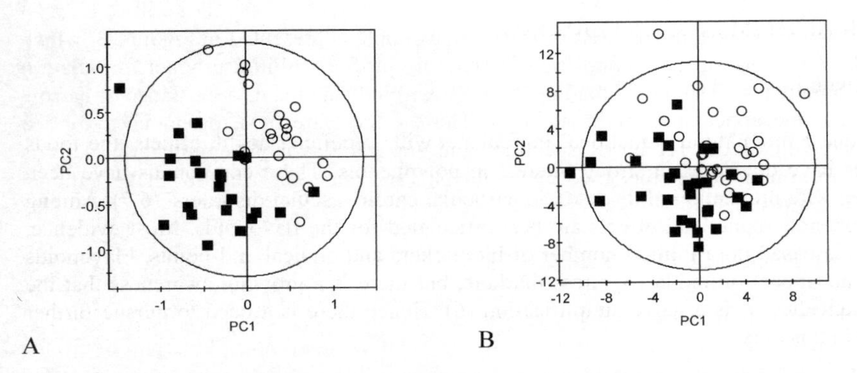

Figure 2 *PLSDA scores plot of (A) urine and (B) plasma ^1H NMR spectra obtained from volunteers on grape/wine extracts (squares) and a placebo (circles).*

2.3 Design/Deploy: rational product formulation

Flavonoids are typically sourced as natural product extracts (NPE's), which have a considerable compositional complexity (24). This poses a challenge for the product developer, since this complicates sourcing of raw materials, reproducible product formulation and also raises regulatory issues with respect to product safety (25). At present, suppliers typically provide results of crude analytical or functional tests, but such data mostly do not relate to product performance, and can also easily be manipulated (26). This is circumvented by recording compositional profiles, where many different compounds are assessed simultaneously, and in an unbiased manner (27). NMR meets these requirements (28), as is illustrated in Figure 3A. Here, a range of commercially available grape/wine extracts were profiled by NMR and represented in a PCA scores plot. One can observe that many extracts cluster in different groups indicating compositional similarity. Such information can be obtained rapidly, and can be used to aid in the selection of raw materials and suppliers in a rational manner.

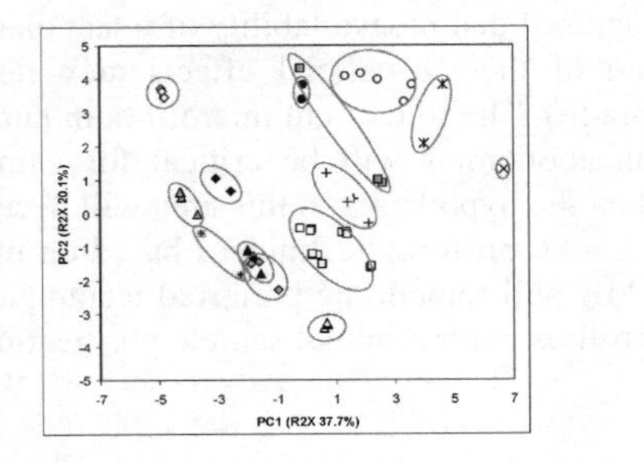

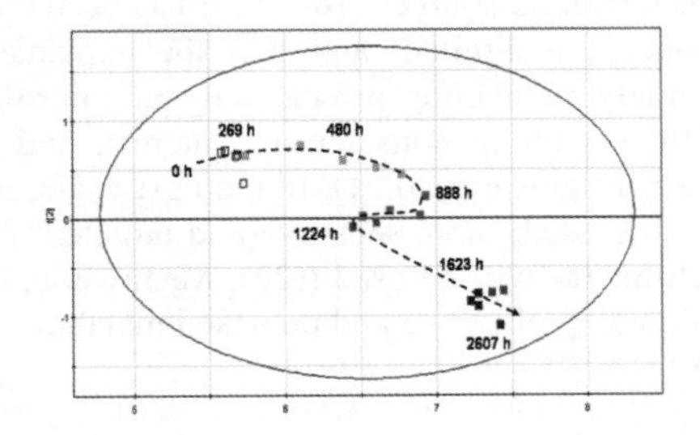

Figure 3 *(A) ^1H NMR based PCA compositional map of commercially available grape extracts. (B) Result of a storage test where the flavonoid composition within a product format was monitored in time by means of ^1H NMR profiles.*

The compositional complexity of NPE's also makes it difficult to make predictions of their stability within formulated products. Also this issue can be addressed by acquiring comprehensive compositional NMR profiles. An example is presented in Figure 3B, which shows the trajectory of the compositional flavonoid profile within a product formulation during a storage test. The model loadings (not shown) indicate the disappearance of narrow signals, and the appearance of broad ones. These effects are most pronounced for the aromatic region of the 1H NMR spectrum and suggest flavonoid aggregation. This is valuable information for the food technologist, and can be used to make rational formulation adjustments.

3 NOVEL FOOD MICROSTRUCTURES

3.1 Food Microstructures-Property relations

Food quality is generally considered to be related to composition, but often the dominating factor is the microstructure of the product (29). Product innovations are often hampered by the lack of understanding of the relation between sensory/physical parameters and the underlying microstructures. The current 'deductive' research strategy in this area is depicted in Figure 4. In time-consuming first step, high-end measurement techniques (imaging, spectroscopy) are deployed, in order to derive quantitative structure descriptors. Next, these descriptors need to be related to sensory parameters. Both steps are projects in themselves, and work on time-scales, which often do not match with the required pace of innovation. Similar to the approach adopted in the Discovery of innovative health ingredients (*vide supra*), we now also witness deployment of explorative modelling in the microstructural area (30,31). As a profiling tool mostly NMR relaxometry is used, due to its reputation in probing food microstructures (32,33) .

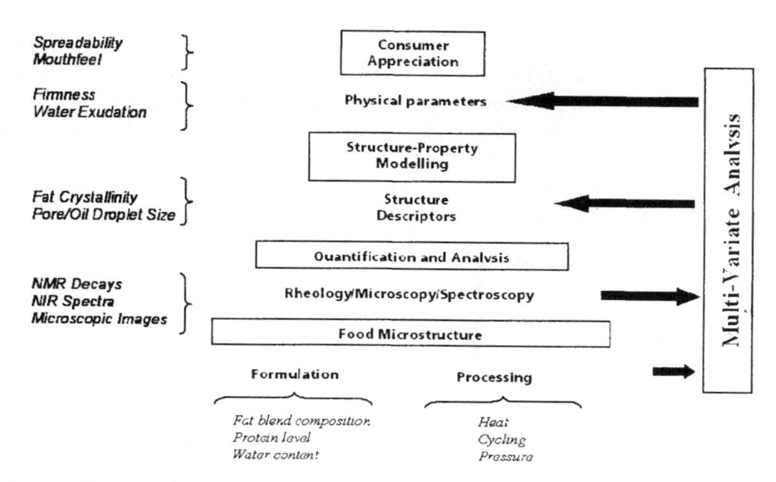

Figure 4 *Structure-property relations established by the 'deductive' rational strategy (pyramid), and by the 'inductive' multi-variate analysis route (right). At the left side examples are given of measurements, structural parameters and, consumer related parameters (bottom to top).*

3.2 Discovery: model emulsion microstructure-property relations

Protein-stabilised oil-in-water emulsions consist of oil droplets stabilised in a protein aggregate network (34) in which water is dispersed in pores with a range of sizes (Figure

5A). An important quality parameter for these food materials is Water Exudation during shelf-life. A range of these emulsions was prepared with varying levels of fat, protein and water, and NMR relaxation decays were recorded to probe their microstructure. These decays contain comprehensive information on microstructure of the food emulsions, which is illustrated in a condensed form in the PCA scores plot in Figure 5B. One can clearly observe clustering, which could be attributed to the presence/absence of a biopolymer.

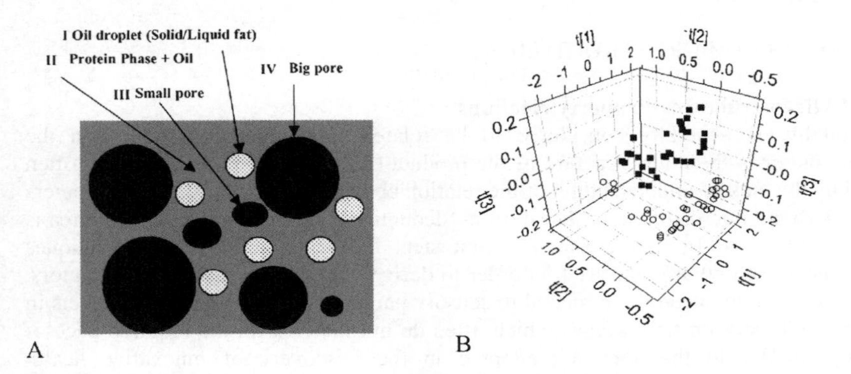

Figure 5 *(A) Schematic representation of microstructures of protein stabilised O/W emulsions. (B) PCA scores plot derived from transversal relaxation decays (^1H NMR, 20 MHz) recorded for a series of these emulsions.*

Figure 6A shows the performance of a multivariate model that predicts Water Exudation from NMR decays recorded for emulsions prepared with biopolymer. (35,36). The model performs poorly for samples without biopolymer, indicating a difference in microstructure. This is visualized in the loadings for models built for samples with/without biopolymer (Figure 6B). The samples without biopolymer have a contribution of small pores, which is absent in samples with biopolymer. This is indicating that the biopolymer had a water retaining effect on the emulsions.

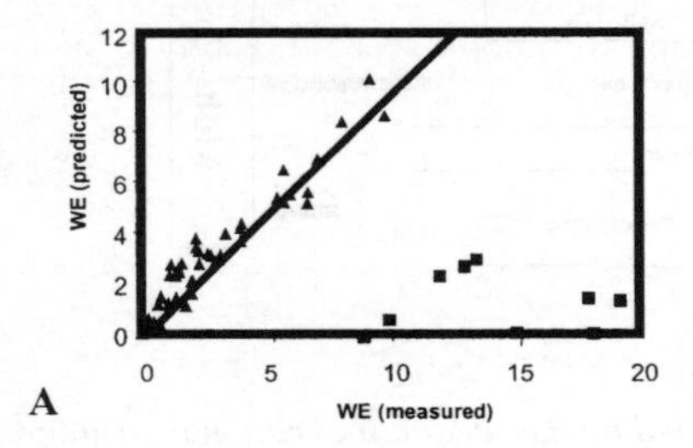

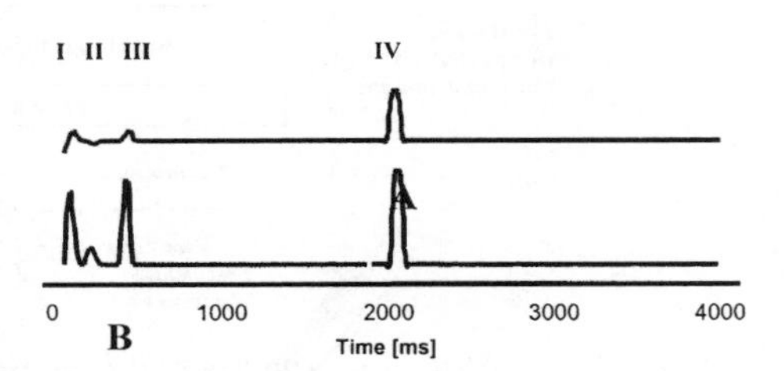

Figure 6 *Goodness of fit for multivariate model based on samples with biopolymer (triangles). In the plot also samples without biopolymer (squares) have been presented (prediction values based on model for samples with biopolymer. (B) T_2 components that explain WE for model based on samples with (top) and without (bottom) biopolymer.*

3.3 Design: rapid and cheap assessment of microstructures

In the Design phase, the pressure to bring a product to the market is increasing, but meanwhile the prototypes under investigation are still complex, both on microstructural and on compositional level. In this stage, food product Designers prefer methods that provide rapid feedback on meso- and microstructure of food product prototypes. Understanding and/or control of food structures at the laboratory bench/kitchen table or manufacturing pilot plant requires relatively cheap and easy-to-use measurement technologies that can be operated by non-(NMR) expert users. Already in the 70's, such systems became available for the routine measurement of Solid Fat Content (37,38) in fat blends, and this rapidly abolished the cumbersome classical method (39). Later, benchtop NMR equipment was extended with the capability to assess water (40,41) and oil (42,43) droplet sizes in food emulsions. Recently, also benchtop NMR methods were presented for assessment of fat /water content (44,45), phase composition and microstructure (46,47) of complex food products.

3.4 Deploy/Deliver: through-package assessment of microstructural quality

In the Deploy and Deliver phases, we are witnessing a transition from the 'classical' process and quality control taking place in an 'off-line' laboratory, towards non-invasive, on-line and real-time measurement of product quality parameters. Most of these measurement tools only provide *chemical compositional* information of the food system of interest, however. Benchtop NMR is well suited for rapid microstructural assessment, but still requires sampling in NMR tubes (48). Recently, NMR has also been presented in a truly non-invasive mode, by deploying one-sided magnets with built-in (49) measurement coils, also denoted as the MObile Universal Surface Explorer (MOUSE). The first applications of the NMR MOUSE were in non-invasive assessment of polymer quality (50), but also the first applications in food technology have appeared. These food applications of the MOUSE focussed on the non-invasive assessment of *compositional* parameters (51,52). However, the aforementioned sensitivity of NMR to distribution and dynamic state of water implies that the MOUSE should also be a versatile sensor for the *microstructural* quality of foods (53). We have explored this in a recent study where we investigated whether the microstructural quality of the aforementioned model emulsions could be assessed by the NMR MOUSE. As a measure for microstructural quality, Water Exudation (WE) was taken, which is commonly assessed by a cumbersome and destructive gravimetric procedure.

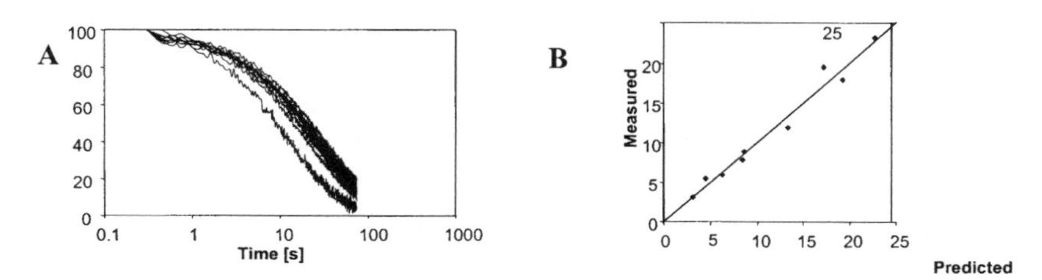

Figure 7 *(A) MOUSE decays of food emulsions that were used for non-invasive (though-package) predictions of water-exudation (B).*

In Figure 7A the NMR decays are presented of a series of emulsions with different WE values. By means of multi-variate methods, models can be built that correlate these decays to WE values (Figure 7B). Validation of these models indicated that the MOUSE yielded reliable 'through-package' measurements of WE (54). This opens opportunities for non-invasive and on-line testing of the microstructural quality of food products in manufacturing environments and in the supply chain.

4 CONCLUSIONS AND PERSPECTIVES

In the Discovery phase, NMR is playing an important role in holistic assessment of complex materials and organisms like humans. The most impressive example is metabonomics, which has profited significantly from the stability and unbiased and comprehensive nature of NMR. These characteristics can also be exploited in the explorative building of understanding of the microstructural origin of the properties of food products and materials. We expect that NMR will keep playing an essential role in the systematic and unbiased building of both physical and biological understanding. For the sake for conciseness we have not been able to address the developments in Discovery of the underlying neurophysiological factors of consumer behavioural/perception where functional MRI (55) has rapidly acquired an essential position.

We have also illustrated how NMR is being applied in the Design/Deploy phases. Implementation of relaxometric and diffusometric NMR methods on benchtop instrumentation, has brought NMR close to the workbench of the food technologists and product developer. We envisage that further miniaturisation of NMR instrumentation will take place in the next years, thus moving NMR technology further into the Deploy and Deliver phase. The MOUSE sensor may only present a first example of a revolution that may profit from developments in magnet design (56) and microcoils (57). These hardware developments may enable truly online sensing (58,59) of meso/micro/macro-structural product quality. Ultimately, we may even envisage the use of hand-held sensors for consumer (health) self measurement (60).

Acknowledgements

The MOUSE work was carried out in collaboration with Dr. V. Litvinov and Dr. D. van Dusschoten (DSM Research).

References

1. WHO/FAO "Diet, nutrition and the prevention of chronic diseases: report of a joint WHO/FAO expert consultation"; WHO/FAO, 2002, Geneva, Switzerland: 2002.
2. Vischer, T. L. S.; Seidel, J. C. *Annual Rev.Publ.Health* **2001,** *22* 355-375.
3. Norton, I.; Fryer, P.; Moore, S. *Aiche Journal* **2006,** *52*(5), 1632-1640.
4. Kell, D. B. *Current Opinion in Microbiology* **2004,** *7*(3), 296-307.
5. Kell, D. B.; Oliver, S. G. *Bioessays* **2004,** *26*(1), 99-105.
6. Scalbert, A.; Johnson, I. T.; Saltmarsh, M. *American Journal of Clinical Nutrition* **2005,** *81*(1), 215S-2217.

7. Williamson, G.; Manach, C. *American Journal of Clinical Nutrition* **2005,** *81*(1), 243S-255.

8. Verrips, C. T.; Warmoeskerken, M. M. C. G.; Post, J. A. *Current Opinion in Biotechnology* **2001,** *12*(5), 483-487.

9. Lindon, J. C.; Nicholson, J. K.; Holmes, E.; Everett, J. R. *Concepts in Magnetic Resonance* **2000,** *12*(5), 289-320.

10. Lindon, J. C.; Holmes, E.; Nicholson, J. K. *Progress in Nuclear Magnetic Resonance Spectroscopy* **2001,** *39*(1), 1-40.

11. Alam, T. M.; Alam, M. K. *Annual Reports on NMR Spectroscopy,* **2005,** *54* 41-80.

12. German, J. B.; Watkins, S. M. *Trends in Food Science & Technology* **2004,** *15*(11), 541-549.

13. Whitfield, P. D.; German, A. J.; Noble, P. J. M. *British Journal of Nutrition* **2004,** *92*(4), 549-555.

14. Watkins, S. M.; Hammock, B. D.; Newman, J. W.; German, J. B. *American Journal of Clinical Nutrition* **2001,** *74*(3), 283-286.

15. Nicholson, J. K.; Holmes, E.; Lindon, J. C.; Wilson, I. D. *Nature Biotechnology* **2004,** *22*(10), 1268-1274.

16. Solanky,K.; Bailey, N.,; Beckwith-hall, C.; Bingham, A.; Davis, A.; Holmes, E.; Nicholson, J.; Cassidy, A. *J Nutrit. Biochem.* **2005,** *16,* 236-244.

17. Wang, Y. L.; Tang, H. R.; Nicholson, J. K.; Hylands, P. J.; Sampson, J.; Holmes, E. *Journal of Agricultural and Food Chemistry* **2005,** *53*(2), 191-196.

18. Daykin, C. A.; Van Duynhoven, J. P. M.; Groenewegen, A.; Dachtler, M.; Van Amelsvoort, J. M. M.; Mulder, T. P. J. *Journal of Agricultural and Food Chemistry* **2005,** *53*(5), 1428-1434.

19. van Dorsten, F.; Daykin, C. A.; Mulder, T. P. J.; Van Duynhoven, J. P. M. *Journal of Agricultural and Food Chemistry* **2006,** *in press.*

20. Nicholson, J. *Drug Metabolism Reviews* **2005,** *37* 10.

21. Nicholson, J. K.; Lindon, J. C.; Holmes, E. *Xenobiotica* **1999,** *29*(11), 1181-1189.

22. van der Greef, J.; Davidov, E.; Verheij, E.; Vogels, J.; van der Heijden, R.; Adourian, A. S.; Oresic, M.; Marple, E. W.; Naylor, S. The role of metabolomics in System Biology, in *Metabolic profiling*, Kluwer Academic Publishers: 2003; pp. 199-216.

23. Pelczer, I. *Current Opinion in Drug Discovery & Development* **2005,** *8*(1), 127-133.

24. Cheynier, V. *American Journal of Clinical Nutrition* **2005,** *81*(1), 223S-229S.

25. Rochfort, S. *Journal of Natural Products* **2005,** *68*(12), 1813-1820.

26. *Food Authenthicity and Traceability;* Woodhead Publishing: 2003.

27. Ashurst, P. R.; Dennis, M. J. *Analytical methods of food authentication;* Blackie Academic & Professional: 2006.

28. Le Gall, G.; Colquhoun, I. J. NMR spectroscopy in food authentication, pp 131-155, in *Food Authenticity and Traceability, ed. M. Lees*, Woodhead Publishing Ltd, Cambridge, 2003: 2005.

29. Norton, I.; Fryer, P.; Moore, S. *Aiche Journal* **2006,** *52*(5), 1632-1640.

30. Munck, L.; Norgaard, L.; Engelsen, S. B.; Bro, R.; Andersson, C. A. *Chemometrics and Intelligent Laboratory Systems* **1998,** *44*(1-2), 31-60.

31. Bro, R.; van den Berg, F.; Thybo, A.; Andersen, C. M.; Jorgensen, B. M.; Andersen, H. *Trends in Food Science & Technology* **2002,** *13*(6-7), 235-244.

32. Thybo, A. K.; Karlsson, A. H.; Bertram, H. C.; Andersen, H. J.; Szcypinski.P.M.; Donstrop, S. NMR and MRI in texture measurement, in *Texture in Food 2: Solid Foods*, Woodhead Publishing Ltd.: 2004; pp. 184-200.

33. Micklander, E.; Thygesen, L. G.; Pedersen, H. T.; van den Berg, F.; Bro, R.; Rutledge, D. N.; Engelsen, S. B. *Multivariate analysis of time domain nmr signals in relation to food quality*; 2003; pp. 239-254.

34. Kiokias, S.; Reiffers-Magnani, C. K.; Bot, A. *Journal of Agricultural and Food Chemistry* **2004,** *52*(12), 3823-3830.

35. Haiduc, A. M.; van Duynhoven, J. P. M.; Heussen, P.; Reszka, A. A.; Reiffers-Magnani, C. K. *Food Research International* **2006,** *in press*.

36. Haiduc, A. M.; van Duynhoven, J. *Magnetic Resonance Imaging* **2005,** *23*(2), 343-345.

37. van Duynhoven, J. P. M.; Goudappel, G. J. W.; Gribnau, M.; Shukla, V. K. S. *AOCS Inform* **1999,** *10*(95), 479-484.

38. van Putte, K.; van den, E. J. *Journal of the American Oil Chemists Society* **1973,** *51* 318-320.

39. Geurtz, T. J. H.; Waddington, D. *Journal of the American Oil Chemists Society* **1980,** *57*(2), A150.

40. Van den Enden, J. C.; Waddington, D.; Vanaalst, H.; Vankralingen, C. G.; Packer, K. J. *Journal of Colloid and Interface Science* **1990,** *140*(1), 105-113.

41. Balinov, B.; Soderman, O.; Warnheim, T. *Journal of the American Oil Chemists Society* **1994,** *71*(5), 513-518.

42. Goudappel, G. J. W.; van Duynhoven, J. P. M.; Mooren, M. M. W. *Journal of Colloid and Interface Science* **2001,** *239*(2), 535-542.

43. van Duynhoven, J. P. M.; Goudappel, G. J. W.; van Dalen, G.; van Bruggen, P. C.; Blonk, J. C. G.; Eijkelenboom, A. P. A. M. *Magnetic Resonance in Chemistry* **2002,** *40* S51-S59.

44. Sorland, G. H.; Larsen, P. M.; Lundby, F.; Rudi, A. P.; Guiheneuf, T. *Meat Science* **2004,** *66*(3), 543-550.

45. Todt, H.; Guthausen, G.; Burk, W.; Schmalbein, D.; Kamlowski, A. *Food Chemistry* **2006,** *96*(3), 436-440.

46. Mariette, F.; Lucas, T. *Journal of Agricultural and Food Chemistry* **2005,** *53*(5), 1317-1327.

47. Bertram, H. C.; Wiking, L.; Nielsen, J. H.; Andersen, H. J. *International Dairy Journal* **2005,** *15*(10), 1056-1063.

48. Nordon, A.; McGill, C. A.; Littlejohn, D. *Analyst* **2001,** *126*(2), 260-272.

49. Blumich, B.; Blumler, P.; Eidmann, G.; Guthausen, A.; Haken, R.; Schmitz, U.; Saito, K.; Zimmer, G. *Magnetic Resonance Imaging* **1998,** *16*(5-6), 479-484.

50. Guthausen, G.; Todt, H.; Burk, W.; Kamlowski, A.; Schmalbein, D. *Kautschuk Gummi Kunststoffe* **2003,** *56*(11), 578-581.

51. Guthausen, A.; Guthausen, G.; Kamlowski, A.; Todt, H.; Burk, W.; Schmalbein, D. *Journal of the American Oil Chemists Society* **2004,** *81*(8), 727-731.

52. Pedersen, H. T.; Ablett, S.; Martin, D. R.; Mallett, M. J. D.; Engelsen, S. B. *Journal of Magnetic Resonance* **2003,** *165*(1), 49-58.

53. Martinez, I.; Aursand, M.; Erikson, U.; Singstad, T. E.; Veliyulin, E.; van der Zwaag, C. *Trends in Food Science & Technology* **2003,** *14*(12), 489-498.

54. Haiduc, A. M.; Trezza, E.; van Dusschoten, D.; Reszka, A. A.; van Duynhoven, J. P. M. *Lebensmittel-Wissenschaft Und-Technologie-Food Science and Technology* **2006,** *in press.*

55. Rolls, E. *fMRI and the sensory perception of food*; In: Magnetic Resonance in Food Science Royal Society of Chemistry: 2005.

56. Hills, B. P.; Wright, K. M.; Gillies, D. G. *Journal of Magnetic Resonance* **2005,** *175*(2), 336-339.

57. Webb, A. G. *Journal of Pharmaceutical and Biomedical Analysis* **2005,** *38*(5), 892-903.

58. Hills, B. P.; Wright, K. M. *Towards on-line NMR sensors*; In: Magnetic Resonance in Food Science, Royal Society of Chemistry: 2005; pp. 175-185.

59. Hills, B. P. Nuclear Magnetic Resonance Imaging, in *Detecting foreign bodies in food*, Woodhead Publishing: Cambrdige, UK, 2004; pp. 154-171.

60. Price, K. E.; Vandaveer, S. S.; Lunte, C. E.; Larive, C. K. *Journal of Pharmaceutical and Biomedical Analysis* **2005,** *38*(5), 904-909.

Food in the Human Body

FUNCTIONAL MRI OF FOOD IN THE GASTROINTESTINAL TRACT

E. Cox[1], C.L. Hoad[1], L. Marciani[2], R.C. Spiller[2] and P. A. Gowland[1]

[1]Sir Peter Mansfield Magnetic Resonance Centre, School of Physics and Astronomy, University of Nottingham, Nottingham, UK
[2]Wolfson Digestive Diseases Centre, Nottingham University Hospital, University of Nottingham, Nottingham, UK

1 BACKGROUND

Thus far Magnetic Resonance Imaging (MRI) has been developed mainly as a tool for clinical diagnosis. However MRI can study function as well as anatomy and since it does not involve the use of any harmful ionizing radiation it can provide a unique tool for making non invasive measurements of normal physiological function in humans. Over the last decade MRI has been used in this way to study the handling of food in the stomach and intestine, and also the brain's response to taste of food. In particular MRI has been used to study the effect of meal viscosity and fat content on the handling of food in the gut. MRI can measure not only movement and volumes but also flow and shear rates and parameters related to viscosity and fat content.

The stomach is a major site for the processing of food, being responsible for its mechanical grinding, dilution, chemical breakdown, mixing and transport to the small intestine. The intestine is responsible for further chemical and mechanical processing prior to absorption. The function of the gastro-intestinal (GI) tract is partly controlled by the brain, but conversely the GI tract provides feedback signals to the brain, and this loop is known as the gut brain axis [1]. In particular the function of the GI tract has a crucial role in determining the sense of satiety [2, 3]. With the increasing trend towards obesity in the developed world, it is essential to understand how the form and nature of a meal affects GI function and also the sense of satiety. The factors affecting satiety can be broken down into cephalic effects (the perceived texture and flavour of food in response to oral stimulation), gastric effects (viscosity and texture, gastric distension, emulsion stability and fat distribution) and intestinal effects (the nutrient delivery profile, entero-endocrine signaling and intestino-intestinal reflexes.

Many reports of gastrointestinal function in the literature are based on invasive animal models, since until recently the methods available to study GI function in humans had various limitations. Naso-gastric intubation is a simple approach, but it is highly invasive and is limited to studying liquid meals that can easily be aspirated. Intraluminal manometry also requires intubation, and as such it is also highly invasive. It can be used to evaluate antral motility, but it can fail to detect non-occlusive contractions. Ultrasound has particular application in easuring intragastric flow velocities. However the ultrasound beam is disrupted by air/liquid interfaces, which usually prevents imaging of the complete gastric contents, and limits its application in the intestine. Gamma scintigraphy is particularly useful for monitoring multiphase meals, by radioactively labeling different food components with different isotopes. However this technique yields poor resolution and cannot detect any secretions added to the digesta. Radiological techniques were used in the past to follow the emptying of radio-opaque markers. They generally have good spatial resolution but have to use non-physiologic meals for adequate contrast and give high radiation doses to subjects. Electrical Impedance Tomography has also been used to monitor gastric emptying. It is highly non-invasive and cheap, but it has limited spatial resolution or contrast to meal distribution.

Image artifacts have limited the use of MRI in the assessment of GI physiology in the past. These arise mainly from the contractions of the gastric wall and breathing contractions of the diaphragm. In recent years there have been promising advances in the use of fast MRI sequences which largely overcome this problem. Work carried out in Nottingham showed that echo-planar magnetic resonance imaging (EPI) at low field (0.5 T) is well suited for dynamically assessing gastric physiology [4]. EPI acquires images very rapidly (in under 130 ms) overcoming artefacts due to abdominal motion. Furthermore, the EPI pulse sequence gives intrinsically T_2* weighted images, and hence provides excellent contrast between water-based liquid meals and surrounding tissues, and between the liquid and solid components of a meal [5, 6], without the need for the addition of MRI contrast agents. EPI can be particularly conveniently combined with other NMR pulse sequences for quantifying various physical parameters. More recently we have also started to use other ultrafast imaging sequences including the fast spin echo sequences such as HASTE, FISP and spoiled gradient echo sequences such as FLASH.

This paper presents the different MRI methods used to study GI function and some of the results we have obtained over the last decade.

2 GENERAL METHODS

This section will describe the measurements that can be made using MRI. We believe that MRI can now be considered to be a 'one stop shop' for the assessment of GI physiology and certainly pathological gastric function in patients.

Subject related issues

Since we are aiming to study normal physiology we generally exclude subjects we a history of gastrointestinal disease, or an abnormal body mass index. We ask subjects to fast overnight before each experiment, and this can be confirmed by an initial scan of the gastric lumen. At the beginning of their first experimental session, subjects are instructed on how to hold their breath before each image acquisition, in order to minimise changes in diaphragmatic displacement. A typical study session may last several hours as this is the time scale of gastric processing of a meal, and data acquisition occurs at approximately 20 to 45 minute intervals, depending on the study. The subjects have to lie down horizontally for the period necessary to acquire the images, but in between each scanning period (usually lsting only a few minutes), the subjects are made to sit upright. If the meal contains materials which we expect to layer because of their different buoyancies, then we might ask subjects to lie with their right side down, to expose the duondenum to the denser component of the meal first (before the less dense component) which would be the normal situation when a subject is sitting upright. If we wish to delay gastric emptying for some reason (for instance to observe the effect of gastric juices on a meal over an extended period of time) we will give a 'fat preload' (i.e. we ask the subjects to ingest approximately 20 ml of fat usually in the form of a nutritional supplement), which will trigger duodenal feedback to slow gastric emptying.

Subjective assessment of the sense of satiety

The sense of satiety is clearly a key factor in human eating behaviour. Although psychological and social factors are undoubtedly important in determining the eating patterns, there are also biological factors that are known to influence how much food is consumed at any particular meal [7]. One example is the food calorie content, which plays an important role in determining satiety as demonstrated by duodenal fat infusion studies [8]. The sense of satiety is also linked to the physical properties of food [9]. Human studies have suggested that viscous meals are more satisfying [10, 11]. This is possibly due to antral stretch receptors, which induce a sense of satiety in response to distension. However to investigating in detail the interactions between the properties of the meal within the gastric lumen, the sense of satiety, and gastric function in humans remains difficult without a non-invasive technique for investigating the relevant parameters, such as MRI.

In most of our studies the volunteers have scored their subjective feelings of fullness, hunger and appetite, each on a scale of 1 -10, at various time points throughout the experiment. We have subsequently compared the area under the curves of each score against time, with the observed changes in gastric function.

MRI pulse sequences

The following sequences have been used:

- Echo planar imaging on a whole-body 0.5 T purpose-built EPI scanner. equipped with actively shielded gradient coils and a 50 cm diameter linear, bird-cage, RF coil. (In-plane resolution 3.5 mm × 2.5 mm, slice thickness of 1 cm, ETL= 130 ms, effective TE= 40 ms.). This sequence is used to measure volumes of the gastric lumen, and can be repeated at around 0.3 Hz if dynamic behaviour is being studied. Inversion recovery EPI and spin echo EPI have also been used to measure the NMR relaxation times of the GI contents. The pulsed gradient spin echo (PGSE) EPI sequence has been used to measure flow rates in the gastric lumen.
- Transverse Balanced TFE *(TrueFISP)* sequence (TE=1.2 ms, TR=2.4 ms, Flip Angle 45°, 20 contiguous slices 10mm thick, scan time = 9s) on a 1.5 T Philips Achieva scanner. This can be used to measure volumes of the gastric lumen, and can also be used to measure the relaxation times of the GI contents.
- Coronal TSE *(MRCP)* on a 1.5 T Philips Achieva scanner (TE=320 ms, TR= 8000 ms, Fat sat SPIR, 24 contiguous slices 7mm thick, scan time 24 s). This has been used to assess water volumes in the small bowel

- Coronal Dual Echo FFE *(gradient echo)* on a 1.5 T Philips Achieva scanner. (TE=2.3 ms and 4.6 ms, TR=156 ms, Flip Angle 80°, 24 contiguous slices 10mm thick, scan time 13 s). This is used to measure the volume of the colon.

3 MEASUREMENTS

Gastric emptying

The rate of gastric emptying of a meal is influenced by many factors [12], including meal pH, osmolarity, calorie content, lipid content, volume, viscosity, mechanical strength and spatial distribution of food components, which can all affect the response of the stomach. In particular the powerful feedback response of the duodenal chemoreceptors, mediated by neurones and hormones particularly cholecystokinin (CCK) [13], influences gastric emptying. Several experiments in animals have proved that meal viscosity is a key factor influencing gastric function [14, 15]. Similarly, human studies have suggested that meal viscosity might slow gastric emptying [16]. However the degree to which these observed changes are a simply a passive result of changes in the physical properties of the gastric contents has proved difficult to study in humans using conventional modalities. We have used EPI to investigate the changes induced in the rate of gastric emptying by varying the physical properties of food. From the rapid, multi-slice, volume data sets it was also possible to separate antral and fundal volumes, as well as to investigate gastric accommodation, by measuring the circumference of the antrum.

Fate of multiphase meals

It is important to be able to assess the intragastric distribution of the liquid, viscous and solid components of a meal, in order to understand of the process of digestion. One approach has been to use dual-energy, gamma scintigraphy techniques, separately labelling the different components of the meal. However the different proton densities and relaxation times of the liquid, viscous and solid components of a meal, make it possible to monitor the fate of each phase separately, serially, accurately and non-invasively using EPI. Furthermore, the chemical shift between the resonances of the water and the fat protons can be exploited to image fat and water components separately by suppressing the signal arising from the other component.

We have used MRI to investigate the intragastric mixing and the gastric emptying of multi-phase meals consisting of oil and water, porridge and water, model solid particulates in a model meal matrix and nutrient and non-nutrient viscous polysaccharides. We have studied the intragastric layering of a mixed-phase fatty meal and used NMR relaxometry to study the fat content of a meal at 0.5 T [17-19].

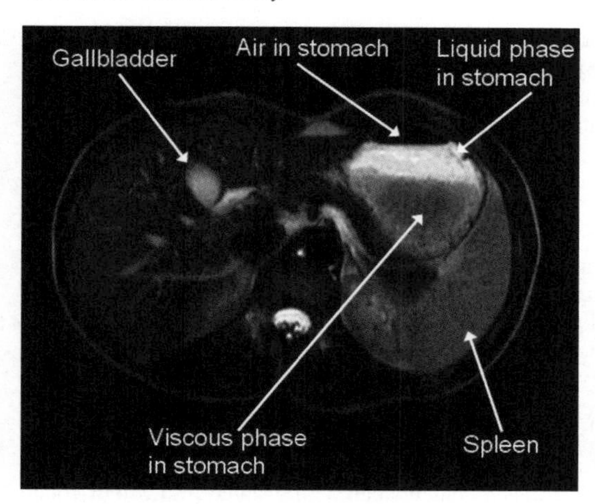

Figure 1: This transverse Balanced TFE *(TrueFISP)* image shows intragastric layering of liquid (bright) above a creamed rice pudding meal in the stomach of a healthy volunteer. A pocket of air is also visible on the top of the stomach.

Intragastric meal viscosity

Several animal studies have indicated that meal viscosity is a key physical factor in influencing gastric motor function [15]. However the techniques to measure intra-gastric viscosity have been invasive, and it is not clear that the results can be directly applied to humans under normal conditions. Human studies have suggested that increased viscosity will slow gastric emptying [20, 21] and that high viscosity meals increase satiety [10]. It has also been proposed that increased meal viscosity might slow absorption of nutrients, and thus improve the management of glucose intolerance and obesity [22-24].

We have developed and validated a technique to measure the viscosity of model polysaccharide test meals in vivo within the gastric lumen [25, 26], by calibrating the transverse relaxation rate (T_2^{-1}) of a simple polysaccharide model meal against its viscosity in vitro. Subsequently we validated the method in vivo by aspirating samples of the gastric contents, and measuring the viscosity of the samples, in vitro, using a viscometer. The EPI *in vivo* viscosity measurements correlated well with aspirates' viscosity. We have used this technique to assess the effect of meal viscosity and nutrient content on intragastric dilution, gastric emptying and motility. We found that the stomach responded to ingestion of a viscous meal by rapid intragastric dilution leading to a reduction in viscosity, which was greatest for the most viscous meal. It was found that a 1000 fold increase in the initial meal viscosity induced only a small delay in gastric emptying. If emptying were purely passive it would be expected that a more viscous meal would be emptied more slowly. Increasing meal viscosity did not induce any changes in the frequency and velocity of antral contractions, suggesting that the force of gastric contractions had increased. Maybe most importantly it was confirmed that increased meal viscosity was associated with increased satiety; this was probably due to the response of stretch receptors in the antrum. In a separate study using model solid meals [27], it was found that increasing meal viscosity impaired gastric sieving of model solid food, that is to say the model meal was emptied more rapidly when the matrix of the meal was more viscous.

Using a series of nutrient and non-nutrient polysaccharide meals it was found that the presence of nutrients delayed gastric emptying, having a stronger effect than meal viscosity.

Interestingly, we demonstrated a linear relationship between gastric volumes and the subjective sense of satiety with low and high viscosity, nutrient meals. We also observed that the volume of secretions within the stomach at one hour after ingestion was higher for the nutrient meal than for the non-nutrient meal. The nutrient contents of the meals did not affect the frequency and velocity of antral contractions.

Intragastric meal dilution and mixing

As it is clear that meal viscosity is an important factor in determining the sense of satiety, it is interesting to study the processes of intragastric meal dilution and mixing. These processes are particularly important for understanding digestion, which is often modelled in vitro assuming that the meal is rapidly exposed to a very acidic environment. For instance this has implications understanding the process of dietary fat hydrolysation, as several oral enzymes are inactivated at low pH when the bolus is mixed with acid gastric secretions.

In vitro calibration curves of T_2^{-1} versus model meal dilution can be used with gastric emptying curves combined with vivo T_2 measurements to calculate meal dilution in vivo [28]. Furthermore pixel-by-pixel T_2 maps of the stomach contents can be translated into colour-coded meal dilution maps using *in vitro* calibrations [29]. By combining the gastric volumes and dilution measurements we have been able to measure the volume of secretions present within the meal with time. One hour after ingestion the volume of secretions was greater for the more viscous meal. We investigated the details of meal dilution initially for a high nutrient, fatty, porridge meal, followed by water. Analysis of signal intensity ratios in the EPI images showed that the porridge hydrated and mixed slowly, and that the meal remained in two phases for more than 90 min. Subsequently we have mapped meal dilution quantitatively for the different Locust Bean Gum based meals and have shown that the high viscosity meals diluted heterogeneously, whilst low viscosity meals were uniformly diluted. Contrary to the common idea of rapid and complete homogenisation of a meal, gastric contents were generally rather poorly mixed. The meal remains heterogeneous for a long time after digestion and the gastric secretions only poorly penetrate the food bolus. The dilution maps data also suggested that the more dilute, peripheral components of the meal bolus left the stomach selectively by a process of elution.

We have also studied the fate of materials designed to gel in the gastric lumen [30]. We demonstrated gelation, and also floating of alginate gel rafts [31]. We also found that strong-gelling alginate increased the sense of fullness as much as a meal that was viscous on ingestion. This suggests that agents that gel on contact with acid may be useful additions to weight reducing diets. This effect may be due to either distension in the gastric antrum or altered transport of nutrients to the small intestine in the gel.

A multiphase aqueous and oil meal has been used to measure the emptying rates of oil and water components separately and to observe the mechanism by which posture alters gastric emptying [32]. When the subject was lying with the right side down, the lipid phase was observed to float in the fundus and emptying of the aqueous phase was fast. With the subject lying with the right side up the lipid layer filled the duodenum, where it could stimulate the duodenal receptors, and gastric emptying was significantly slower. In a subsequent experiment we confirmed that fat delays gastric emptying and we were able to assess intragastric flow non-invasively. High-calorie fatty meals were found to be associated with substantial forward and backward flow-events in the antrum, compared to high-calorie, non-fatty meals. Flow was minimal in the gastric body. This antral activity could aid fat emulsification. More recently we have compared the intragastric behaviour of emulsions designed to be stable or cream in the acidic gastric environment. We have shown that fat is emptied slower from the stomach for meals that cream, and we have shown that this can have an effect on the sense of satiety [19].

Antral motility and flow

The conventional method of monitoring gastric motility is to use intraluminal manometry, which is very invasive. However MRI techniques offer the possibility of non-invasively investigating antral motility [33-35]. The antral lumen can be continuously imaged at rapid intervals and the periodic contractions of the antral walls can be detected [34-36]. We have validated this technique against manometry, and have found that more contractions are detected using MRI, which would be consistent with manometry failing to detect the non-occlusive contractions [33]. Such motility plots have been used to investigate the effect of meal viscosity, nutrient and fat content on antral motility.

Antral motility results in the movement of food within the antrum. In the past intragastric flow could only be inferred from the gastric emptying rates, until recent Doppler ultrasound studies measured backward and forward flow events in the antrum. We have shown that the flow of the meal within the antrum can also be assessed using the pulsed gradient spin echo (PGSE) EPI technique [37], and indeed using 'tagging' in which a grid is magnetically superimposed onto the image by spatially modulating the longitudinal magnetisation prior to imaging. Both methods allow the measurement of intragastric velocities and hence flow rates.

Observations of particulates in meals

Although the largest shear rates in the body occur in the mouth, in the post-prandial period, the stomach breaks down meal solids into smaller particles before they can be emptied via the pylorus. This occurs in the antro-pyloric region, where retrograde flow occurs to maximize shear forces. The pyloric sphincter acts as a sieve to impede large particles from leaving the stomach. MRI can easily discriminate between the solid and liquid components of a meal, and therefore antral grinding forces can be measured indirectly using MRI, by monitoring the breakdown and emptying of sets of solid agar gel beads each designed to have a different breakdown force [27]. We have used this technique to estimate the intragastric grinding forces, and the effect of meal viscosity on pyloric sieving of the intragastric contents.

To study the effect of particulates within the meal, we added the model solid particulates to the liquid/viscous polysaccharide meals. In the presence of model solid particles the gastric emptying curves for the total meal volume were exponential and increasing particle fracture strength delayed emptying with both a liquid and a viscous meal. Also, increasing the strength of solid particles significantly increased the sense of fullness for both liquid and viscous nutrient meals. Again, a linear relationship between the sense of fullness and the total gastric volume was found. Hard particulates increased maximum antral diameters but no change was observed in the frequency and velocity of antral contractions.

An indirect measurement of the bead's residence time made it possible to conclude that the maximum grinding force exerted by the human antrum is 0.65 N [27]. Increasing meal viscosity impaired gastric sieving. Softer beads are rapidly broken, and empty at the same rate as a liquid meal. Hard beads increased satiety by a stretch mechanism that involves the antrum.
We have also observed the persistence of beads in the small intestine and have used this to draw conclusions about the efficiency of gastric sieving.

Small bowel water content

Small bowel water content is assessed by segmenting out the gall bladder, major vessels and bladder from MRCP images of the abdomen, and then thresholding the images at a level that removes signal from other tissues such as muscle. The remaining pixels are in the gut and contain free water. The number of these pixels can be used to estimate the small bowel water content. We have used this approach to study the effect of ingestion bran on small intestinal function. We found that including bran in a meal considerably increases small bowel water content.

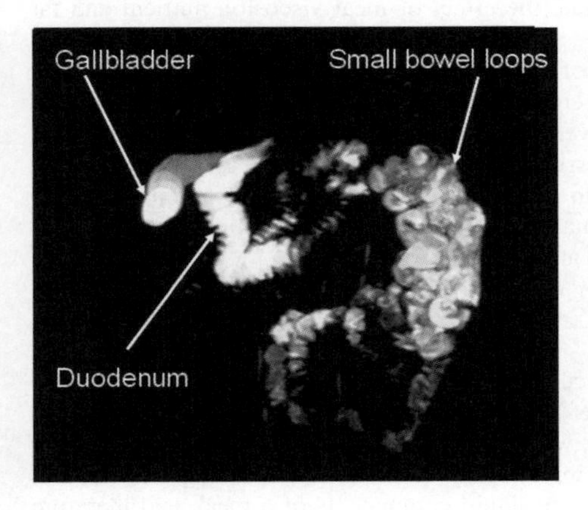

Figure 2: This figure shows a maximum intensity projection of MRCP coronal series acquired across the abdomen of a volunteer 2 hours after feeding. The stomach, bladder and main blood vessels were segmented out. The duodenum and loops of the small bowel can be seen filled with liquid.

Colonic volume

This is measured by outlining the colon on every slice of a gradient echo image. We have only recently introduced this measurement, but have already observed the contraction of the ascending colon 180 minutes after feeding a rice pudding meal. It will be particularly interesting to study small bowel water content and colonic responses in subjects suffering from irritable bowel syndrome.

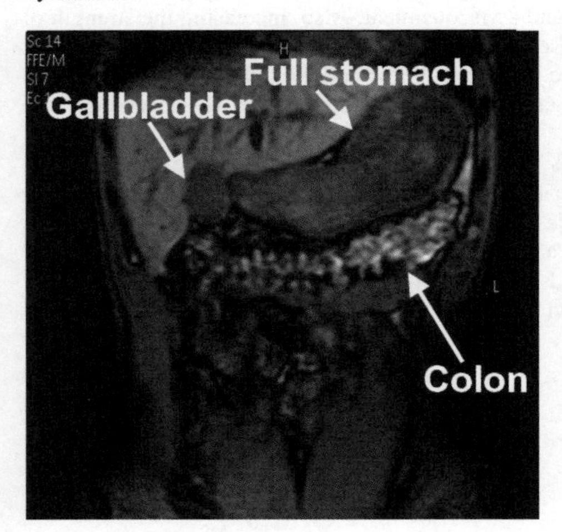

Figure 3: Coronal Dual Echo FFE *(gradient echo)* images of the transverse colon acquired at 1.5 T. The figure on the left is acquired at baseline before feeding and shows the appearance of the transverse colon when the stomach is empty. The figure on the right is acquired immediately after feeding to the volunteer and shows the full stomach.

4 DISCUSSION

This paper has shown that it is possible to assess many different aspects of gastric function in humans using non-invasive MRI techniques. MRI can now provide a range of sequences fast enough to give images devoid of motion artefact, although it should be noted that we have limited our imaging to 1.5T since at high field limits on the rate of RF power may cause an increase in imaging time. A variety of imaging sequences are available with a range of contrasts, making it generally possible to outline the stomach contents easily, and to discriminate between different food components within the digesta, and perform physiological measurements without the need for contrast agent or the use of non-physiological expansion of the GI lumen.
It is still the case that surprisingly little is known about the function of the GI tract and its interaction with the lumen is far from simple and further investigation is certainly required. As the number of MRI scanners increases, it is likely that these investigations will become the preferred method of assessing gastric function, particularly as using MRI it is possible to measure many relevant parameters in a single study. The alternative modalities are generally invasive or involve a radiation dose, and often only measure a single parameter of interest.

We have presented a review of the MRI techniques we use to assess gastrointestinal physiology using MRI. We believe that ultra-fast MRI provides a powerful, non-invasive tool, which is in many ways superior to other commonly used techniques, and which is unique in providing such a wide range on information during a single examination.

ACKNOWLEDGMENTS

We thank the Biotechnology and Biological Sciences Research Council (Swindon, U.K.) for funding most of the work described in this paper. Dr. Annette-Fillery-travis, Dr. martin Wickham, Dr. Phil Boulby, Mr. Richard Faulks, Mr. Paul Young, Dr. Pretima Manoj and Dr. Jeff Wright made considerable contributions to the work reviewed here. Dr. Ron Coxon, Mr. Paul Clark, Ms. Rachel Moore, Mr. Damien Tyler, Dr. Bashar Issa, Ms. Debbie Bush, Ms. Barbara Pick, Dr Jon Hykin and Professor Sir Peter Mansfield also made contributions to the work reviewed.

REFERENCES

1. Aziz, Q. and D.G. Thompson, *Brain-gut axis in health and disease.* Gastroenterology, 1998. **114**(3): p. 559-578.
2. Bray, G.A., *Afferent signals regulating food intake.* Proceedings of the Nutrition Society, 2000. **59**(3): p. 373-384.
3. Berthoud, H.R., *Multiple neural systems controlling food intake and body weight.* Neuroscience And Biobehavioral Reviews, 2002. **26**(4): p. 393-428.
4. Evans, D.F., et al., *Investigation of Gastrointestinal Motility with Echo Planar Magnetic-Resonance Imaging (Epi).* Gut, 1988. **29**(10): p. A1452-A1452.
5. Stehling, M.K., et al., *Gastrointestinal-Tract - Dynamic Mr Studies with Echo-Planar Imaging.* Radiology, 1989. **171**(1): p. 41-46.
6. Evans, D.F., et al., *Prolonged Monitoring of the Upper Gastrointestinal-Tract Using Echo-Planar Magnetic-Resonance-Imaging.* Gut, 1993. **34**(6): p. 848-852.
7. Cecil, J.E., *Oral, gastric and intestinal influences on the control of appetite and feeding in humans.* Appetite, 2001. **36**(3): p. 235-236.
8. Welch, I.M., C.P. Sepple, and N.W. Read, *Comparisons of the Effects on Satiety and Eating Behavior of Infusion of Lipid into the Different Regions of the Small- Intestine.* Gut, 1988. **29**(3): p. 306-311.
9. Santangelo, A., et al., *Physical state of meal affects gastric emptying, cholecystokinin release and satiety.* British Journal of Nutrition, 1998. **80**(6): p. 521-527.
10. Dilorenzo, C., et al., *Pectin Delays Gastric-Emptying and Increases Satiety in Obese Subjects.* Gastroenterology, 1988. **95**(5): p. 1211-1215.
11. Bergmann, J.F., et al., *Correlation between Echographic Gastric-Emptying and Appetite - Influence of Psyllium.* Gut, 1992. **33**(8): p. 1042-1043.
12. Horowitz, M., et al., *Role and Integration of Mechanisms Controlling Gastric-Emptying.* Digestive Diseases and Sciences, 1994. **39**(12): p. S7-S13.
13. Beglinger, C. and L. Degen, *Fat in the intestine as a regulator of appetite - role of CCK.* Physiology & Behavior, 2004. **83**(4): p. 617-621.
14. Prove, J. and H.J. Ehrlein, *Motor Function of Gastric Antrum and Pylorus for Evacuation of Low and High-Viscosity Meals in Dogs.* Gut, 1982. **23**(2): p. 150-156.
15. Cherbut, C., et al., *Action of Guar Gums on the Viscosity of Digestive Contents and on the Gastrointestinal Motor Function in Pigs.* Digestion, 1990. **46**(4): p. 205-213.
16. Sandhu, K.S., et al., *Effect of Pectin on Gastric-Emptying and Gastroduodenal Motility in Normal Subjects.* Gastroenterology, 1987. **92**(2): p. 486-492.
17. Marciani, L., et al., *Fat emulsification measured using NMR transverse relaxation.* Journal of Magnetic Resonance, 2001. **153**(1): p. 1-6.
18. Marciani, L., et al., *Intragastric oil-in-water emulsion fat fraction measured using inversion recovery echo-planar magnetic resonance imaging.* Journal of Food Science, 2004. **69**(6): p. E290-E296.
19. Marciani, L., et al., *Magnetic resonance imaging (MRI) insights into how fat emulsion stability alters gastric emptying.* Gastroenterology, 2003. **124**(4): p. A581-A581.
20. Wilmshurst, P. and J.C.W. Crawley, *The Measurement of Gastric Transit-Time in Obese Subjects Using Na-24 and the Effects of Energy Content and Guar Gum on Gastric-Emptying and Satiety.* British Journal of Nutrition, 1980. **44**(1): p. 1-6.
21. Krotkiewski, M., *Effect of Guar Gum on Body-Weight, Hunger Ratings and Metabolism in Obese Subjects.* British Journal of Nutrition, 1984. **52**(1): p. 97-105.

22. Zavoral, J.H., et al., *The Hypolipidemic Effect of Locust Bean Gum Food-Products in Familial Hypercholesterolemic Adults and Children.* American Journal of Clinical Nutrition, 1983. **38**(2): p. 285-294.
23. Blackburn, N.A., et al., *The Mechanism of Action of Guar Gum in Improving Glucose-Tolerance in Man.* Clinical Science, 1984. **66**(3): p. 329-336.
24. Haskell, W.L., et al., *Role of Water-Soluble Dietary Fiber in the Management of Elevated Plasma-Cholesterol in Healthy-Subjects.* American Journal of Cardiology, 1992. **69**(5): p. 433-439.
25. Marciani, L., et al., *Echo-planar imaging relaxometry to measure the viscosity of a model meal.* Journal of Magnetic Resonance, 1998. **135**(1): p. 82-86.
26. Marciani, L., et al., *Gastric response to increased meal viscosity assessed by echo- planar magnetic resonance imaging in humans.* Journal of Nutrition, 2000. **130**(1): p. 122-127.
27. Marciani, L., et al., *Assessment of antral grinding of a model solid meal with echo- planar imaging.* American Journal of Physiology-Gastrointestinal and Liver Physiology, 2001. **280**(5): p. G844-G849.
28. Marciani, L., et al., *Noninvasive echo-planar imaging (EPI) monitoring of intragastric viscosity, dilution and emptying of viscous meals in normal subjects.* Gastroenterology, 1998. **114**(4): p. G3282.
29. Marciani, L., et al., *Effect of meal viscosity and nutrients on satiety, intragastric dilution, and emptying assessed by MRI.* American Journal of Physiology-Gastrointestinal and Liver Physiology, 2001. **280**(6): p. G1227-G1233.
30. Hoad, C.L., et al., *In vivo imaging of intragastric gelation and its effect on satiety in humans.* Journal of Nutrition, 2004. **134**(9): p. 2293-2300.
31. Marciani, L., et al., *Echo-planar magnetic resonance imaging of Gaviscon alginate rafts in-vivo.* Journal of Pharmacy and Pharmacology, 2002. **54**(10): p. 1351-1356.
32. Boulby, P., et al., *Use of echo planar imaging to demonstrate the effect of posture on the intragastric distribution and emptying of an oil/water meal.* Neurogastroenterology and Motility, 1997. **9**(1): p. 41-47.
33. Wright, J., et al., *Validation of antroduodenal motility measurements made by echo-planar magnetic resonance imaging.* Neurogastroenterology and Motility, 1999. **11**(1): p. 19-25.
34. Marciani, L., et al., *Antral motility measurements by magnetic resonance imaging.* Neurogastroenterology and Motility, 2001. **13**(5): p. 511-518.
35. Schwizer, W., et al., *Measurement of Gastric-Emptying and Gastric-Motility by Magnetic-Resonance-Imaging (MRI).* Digestive Diseases and Sciences, 1994. **39**(12): p. S101-S103.
36. Marciani, L., et al., *Magnetic resonance imaging (MRI) assessment of gastric emptying and antral motility in clinical practice: Preliminary results on patients.* Gastroenterology, 2000. **118**(4): p. 2056.
37. Boulby, P., et al., *Fat delays emptying but increases forward and backward antral flow as assessed by flow-sensitive magnetic resonance imaging.* Neurogastroenterology and Motility, 1999. **11**(1): p. 27-36.

NUTRIMETABONOMICS: METABONOMICS IN FOOD SCIENCE

H. Tang[1] and Y. Wang[2]

[1]State Key Laboratory of Magnetic Resonance and Molecular and Atomic Physics, Wuhan Institute of Physics and Mathematics, The Chinese Academy of Sciences, Wuhan, 430071, PR China.
[2]Chemical and Molecular Systems Biology, Faculty of Medicines, Imperial College London, South Kensington London SW2 7AZ, UK

1 INTRODUCTION

Food science is one of the oldest branches of science and its importance has been increasingly emphasized with the economical and social development. The complexity of food itself has long been realized though such complexity has imposed great difficulties as well for scientists to bring about clear and full molecular understandings to the food composition, the effects of food on the health of consumers, the composition-based food quality assessments and the consumer perceptions. In order to achieve quantitative, universal, integrated and predictive(QUIP) understandings to food and its effects on human biology, technologies with enormous analytical power are required; chemists, biologists, physicists, mathematician, material scientists and food scientist are required to work together in a collaborative and integrative manner. As one of the most important topics in food science, nutritional biochemistry deals with the metabolism of food components, whether they are nutrients and non-nutrients, and the effects of the food components on human endogenous biochemistry upon consumption.

1.1 Metabolism

The word "metabolism" was evolved from the Greek words "*metabole*" meaning change, "*metaballein*" meaning "to change" and the combination of "*meta*" meaning over and "*ballein*" meaning "to throw". Metabolism is defined as the total (bio)chemical changes in living organisms, which includes the biosynthesis of complex organic molecules (anabolism) and their breakdown (catabolism) *via.* sequences of enzymatic steps, known as metabolic pathways. The metabolic activities are thus considered the fundamental characteristics and material basis for all the living systems; the living activities are considered ceased when metabolism stops. Therefore, metabolite analysis has been vital for understanding the molecular basis of living systems.

The importance of metabolism is also highlighted by more than half-a-dozen Nobel prizes awarded to scientists (www.nobel.se) who have made contributions to the understandings of the metabolic mechanisms, such as the metabolism of lactic acid, citric acid (or TCA cycles) and cholesterol. These studies have undoubtedly laid the foundation for the modern biochemistry and enhanced our understandings to the molecular aspects of the most biochemical processes of the living systems. However, these were all based on a reductionism research philosophy, dealing with the metabolic pathways or part of them separately. The living systems rarely work as the sum of individual parts but functioning in a synergistic fashion, thus have to be considered as a holistic entity.

Coincided with dawning of the twenty-first century, there has been a transformation of research philosophy from the reductionism to the holism and the living organisms have been increasingly considered as an integrated complement with each parts functioning in a concerted way. This has led to the emergence of the systems biology concept and explosive occurrence of "omes" and "omics" sciences[1]. With such new concepts coupling with relevant technologies, it is expected that we will be able to achieve quantitative, universal, integrated and predictive (QUIP) understandings to the molecular mechanisms of the living systems[1]. Amongst many, metabonomics is one of the most important new concepts to address the problems in the metabolic level. This is because that whilst genes and proteins provide the fundamental material basis for a biological event to happen (it may not happen!), the metabolic changes represent what actually have happened already with embedded information about the events in the levels of not only metabolism but also proteins and genes. In fact, DNA/RNA and proteins are also metabolites.

1.2 Metabonomics and metabolomics

Metabonomics is also originated from the Greek words *metabole* for "*change*" and *nomos* for "law" (as in economics), thus is actually not an "omics" term but "nomics" one. It was originally defined as[2,3] "*the quantitative measurement of the multi-parametric metabolic response of living systems to pathophysiological stimuli or genetic modifications*" although it was sometimes mistakenly considered as a term used only in toxicity context. The definition clearly emphasizes the integrated and dynamic nature of the living systems.

There is another term "metabolomics" which was coined sometime after the birth of metabonomics. At least half-dozen definitions have been found for metabolomics, such as "*systematic study of the unique chemical fingerprints that specific cellular processes leave behind*"[4], "*the comprehensive analysis of large numbers of metabolites*" (www.nobel.org), "*a high-throughput technology providing the most comprehensive, non-biased analysis of complex metabolite mixtures typical of plant extracts*" (www.plant.wur.nl). The other definitions for it include "*the quantitative measurement of all low molecular weight metabolites in a given sample, cell tissue and the integration of the data in the context of gene function analysis*" (www.metabolomics.bbsrc.ac.uk), "*the quantitative measurement of all low molecular weight metabolites in an organism's cells at a specified time under specific environmental conditions*" (www.jic.bbsrc.ac.uk) and "*comprehensive analysis of all the metabolites of an organism*"[5]. However, the "smallness" of the metabolites in these metabolomics definitions has not been addressed. Nevertheless, as the subject is still in its infancy, the definition is still undergoing evolution. Recently, for example, a new term "*dynamic metabolomics*" was offered with similar meanings to that of metabonomics[1].

It is conceivable that with more discussions and further development, two terms may converge. In the future, metabonomics/metabolomics may be defined[1] as "the branch of science concerned with the quantitative understandings of the metabolite complement of integrated living systems and its dynamic responses to the changes of both endogenous factors (such as physiology and development) and exogenous factors (such as environmental factors and xenobiotics)". As a holistic approach, metabonomics detects, quantifies and catalogues the time related metabolic processes of an integrated biological system, ultimately, relates such processes to the trajectories of the physiological and pathophysiological events[1]. Among many exogenous factors, nutrients and non-nutrients in food have important impacts on the biochemistry of human and associated microflora[1-3].

1.3 Analytical technologies for metabonomics

Since metabolite complement of biological systems are complex mixtures of organic metabolites with various polarity and inorganic salts, metabolite detection, identification and quantification are all vitally important. Analysis of complex mixtures requires considerable amount of analytical power from the technologies. So far many analytical techniques and their combinations have been exploited in metabonomics research, which can be broadly classified[1] into chromatography-based, mass spectrometry (MS) based and NMR-based methods. In general, chromatography- and MS-based methods are well developed selective detection methods, offering *biased detection sensitivities* that vary from metabolites to metabolites[1]. Since chromatographic methods are generally based on the interactions between the analytes and stationary materials, no direct molecular structural information are obtainable. Measuring the masses of the given molecule and its fragments, MS methods at best can offer the formula of analytes and are inadequate in unambiguously identifying isobaric molecules, which are particularly abundant and important in the context of biological samples[1]. Furthermore, the unpredictable metabolite ionization behavior (*e.g.*, ion suppression) makes quantification difficult for MS methods.

The combination of those two methods has made some improvements but biased detection, structural identification and quantification remains to be addressed. In principle, the unambiguous metabolite identification or structural determinations using those methods can be achieved with the availability of standards; the metabolite quantification with those two methods can also be achieved with the *in situ* calibration curves constructed using standards. In practice, however, the complexity of biological metabonome makes it tedious to construct calibration curves for so many metabolites and, even worse, not all standards for the metabolites in biological metabonome are easily available. Furthermore, unknown metabolites (or new biomarkers), by default, are not readily available as standards. Moreover, with the requirement of extensive extraction and sometimes chemical derivatization, those methods lead to complete loss of compartmentation information, being almost impossible to conduct *in vivo* studies. Nevertheless, GC-MS and UPLC-MS have enjoyed successes with improved resolution and reproducibility though quantification, structural identification and isomer discrimination remain to be dealt with.

In contrast, NMR-based methods provide the atomic connectivity of the molecules in pure form or in mixtures and offer universal (or unbiased) detection of all metabolites quantitatively. Its non-destructive nature enables compartmentation information to be reserved favoring the *in vivo* studies. So far, this method is also the only one for direct studies of the intact tissue with high resolution magic-angle spinning NMR. Information of molecular dynamics and interactions are also readily obtainable from the NMR studies. Low intrinsic sensitivity remains the drawback of NMR methods though that can be overcome by some simple preparations to increase the concentration as in the cases of chromatography and mass spectrometry. Two other ways of improving NMR sensitivities include cryogenic probe technology and dynamic nuclear polarization techniques. With the newly developed cryogenic probe technologies, it is now possible for NMR to achieve the urinary metabonome detection within one minute. Although NMR hardware is generally still expensive, the efficiency in terms of structural determination, new biomarker discovery, compartmentation information and high throughput make NMR less expensive (ca. 10 USD per sample) when considered on the costs per sample basis.

It is clear that none of the current techniques *alone* will be able to solve all the problems encountered in the metabonomic analysis. The choice of methods will be related to the purposes of research. The combination of different methods is probably the best strategy forward, such as the newly developed HPLC-SPE-CryoNMR-MS technology[6], which takes advantage of the analytical methods in a combined manner.

2 APPLICATIONS OF METABONOMICS IN FOOD SCIENCE

2.1 Nutrimetabonomics and Phytometabonomics

Nutrimetabonomics, or nutritional metabonomics, can be defined as the branch of science concerned with the quantitative understandings of the dynamic responses of mammalian metabonome to the changes of the nutritional status. It has to be noted that the nutritional effects include those on the biochemistry of both host and gut microflora. Since many phytochemicals in foods, such as polyphenols, were not classified into the category of nutrients, a term "phytometabonomics" is also proposed to define the metabonomics concerned with the global composition and effects of the phytochemicals on the metabonome of both the host and micro flora. This term is particularly useful when herbs and phytomedicines are concerned with. Nutrimetabonomics and phytometabonomics can be considered as sub-topics of metabonomics.

2.2 Metabonomics in food quality controls based on the global composition

Quality controls for foodstuffs are always important for both industry and regulatory authorities. Although there are many ways of monitoring food quality in terms of nutritional values, composition and consumers' perception, the global composition of foodstuffs is one of the most comprehensive ways of describing the food holistically.

As a proof-of-principle study, NMR was applied to understand the composition of fruit juices and vinegars[7]; a number of chemical components was assigned unambiguously and the feasibility of using high resolution NMR in studies of the fluid foodstuffs was confirmed. When in conjunction with multivariate data analysis, NMR spectroscopy was effective to discriminate juices from different apple varieties[8,9]. More importantly, with the spectral assignments, the marker compounds were realized. The results revealed great potential of the method not only in the authentication of apples and apple juices in terms of products of origins and treatments, but also in the analysis of biochemical changes occurring in the fruits[8,9]. These have been further exemplified in a study of the potential of NMR spectroscopy and multivariate analysis methods as a reliable adulteration detection method[10] for orange juices. More than twenty NMR signals differed significantly between the authentic and pulp wash spiked groups. These together with dimethylproline were found as marker compounds for orange juice adulteration. The advantages of this method include its rapidity, simplicity, and diversity of information provided.

Furthermore, NMR spectroscopy was used to follow the compositional changes in mango[11] during ripening and ^{1}H and ^{13}C NMR spectra were recorded to follow the changes in mango pulps using HRMAS techniques[11]. Several organic acids, amino acids, and other minor components were identified and the compositional changes upon ripening were followed through by monitoring the changes in the spectra. In pulps, for instance, sucrose was found to be dominant over fructose and glucose at most ripening stages, and the level of citric acid decreased markedly after the initial ripening stages accompanied with the significant rise of alanine level. This work showed that the direct characterization with NMR offers a unique opportunity to study the overall biochemistry in the whole fruit non-invasively. Similarly, a detailed analysis of the proton NMR spectra of tomato juice and pulp has been reported[12] with a catalogue of 2D NMR methods including J-resolved, COSY, TOCSY, DOSY, ^{1}H-^{13}C HSQC and ^{1}H-^{13}C HMBC; the chemical composition of tomato juices from two cultivars (Red Setter and Ciliegino) was determined[12] comprehensively.

Later, NMR-based metabonomics techniques were employed to study the potential unintended effects in genetically modified tomatoes[13] with simultaneous over-expression of the maize transcription factors; the metabolite composition of these genetically modified tomatoes were compared with that of azygous (non-modified) controls grown side-by-side under the same conditions and the metabonomic changes were monitored during the process of ripening. The results showed[13] that the levels of glutamate, fructose, and some nucleosides increased gradually during the ripening process, whereas amino acids such as valine and γ-aminobutyrate had higher concentration in unripe tomatoes and decreased during the ripening process. Apart from the significantly increased content of six kaempferol glycosides, the levels of at least 15 other metabolites were different[13] between the two types of red tomato, such as citric acid, sucrose, phenylalanine, and trigonelline.

In another study, a set of 191 green teas from different countries was collected and analyzed[14] with ^{1}H NMR in conjunction with principal component analysis (PCA) and cluster analysis. Although the results did not allow allocation of samples to individual countries, cluster analysis suggested that it might be possible with an augmented sample set. For example, the PCA did[14] show a separation between the Longjing type Chinese tea and most other Chinese teas. Even more importantly, some marker metabolites responsible for the difference were detected; Longjing teas showed[14] higher levels of theanine, gallic acid, caffeine, epigallocatechin gallate, and epicatechin gallate and lower levels of epigallocatechin when compared with other teas. Although these compounds had been mentioned prior to the NMR study in connection with quality, the NMR study[14] also showed extra findings that higher levels of theogallin (5-galloylquinic acid), 2-O-(β-L-arabinopyranosyl)-*myo*-inositol, theobromine and some minor sugar-containing compounds were found in Longjing teas whilst higher levels of fatty acids and sucrose were found in the other teas. These new markers can be useful for the authentication of bulk tea[14]. Another study demonstrated[15] the powerfulness of the combination of NMR with chemometrics in quality controls of chamomile tea in terms of its origin, purity and processing methods[15]. An excellent example has also been reported[16] for potential quality controls in beer production and authenticity.

2.3 Metabonomic studies of the effects of food on mammalian biochemistry

The other important applications of metabonomics in food science aim to understand the food effects on human biochemistry and the stress management is particularly relevant in this context. Although the stress effects are complex, there seems an association between long term exposure to high levels of stress and development of multiple forms of chronic illness, such as muscle atrophy[17], insulin resistance and diabetes[18], hypercholesterolemia, hyperlipidemia[19], and immune suppression[20,21]. The association is attributed to the stress-induced elevation of hormones since it is known that acute stress results in status changes of hormones, such as glucocorticoids and catecholamines[17-19]. It has also been shown that there is a correlation between the chronic neonatal stress on newborn rats, such as maternal separation, and the vulnerability of rats to develop irritable bowel syndrome[22] later in life. Conventionally, it is believed that certain dietary components may have some ameliorated effects[23]; proper dietary management may be one of the ways to combat the stress-induced problems. For example, it is well documented in scientific literature that the intake of long chain polyunsaturated fatty acids (LCPUFAs) is essential for good health since LCPUFAs are the basic constituents of cell membranes, determining the cellular membrane fluidity and modulating enzyme activities, carriers and receptors. However, LCPUFAs often have to be ingested as part of the diet since they cannot be synthesized *de novo* in the human body. Based on these understandings, it is deduced that the dietary addition of PUFAs can possibly be helpful in preventing or delaying the effects of some chronic diseases[23-25] and neuronal diseases[26].

To substantiate the propositions and to advance understandings on the stress-induced diseases and the beneficial effects of dietary intervention in the molecular level, there is a clear need for more investigative work to be done on the systemic metabolic consequences of stimuli. However, such studies are difficult to be done in human populations with their huge diversity in genetic and environmental backgrounds. In addition, the conventional biochemistry methods are not effective enough to assess the metabolic responses of mammals to stimuli since these approaches are typically time-consuming and fragmentary such as a series of targeted biochemical assays. Nevertheless, such studies have now become possible with some recent development in holistic platforms such as, microarrays, proteomics, and metabonomics.

2.3.1 Biochemistry of rat small intestine[27]. To establish the metabonomic baseline for models, the high resolution magic-angle spinning (HRMAS) ^{1}H NMR spectroscopy combined with the multivariate data analysis (*i.e.* metabonomics technology) has been employed to characterize the biochemistry of rat small intestine[27]. The results showed that the metabolic profiles of intact jejunum and ileum parts of small intestine tissues from male Long Evens rats[27] chiefly consist of various amino acids, lipids, glycerophosphocholine (GPC), choline, creatine, ethanol, carboxylic acids and nucleoside bases such as cytosine and uracil. The overall biochemical differences between jejunum and ileum tissues can be readily distinguished with principal component analysis (PCA) on their HRMAS ^{1}H NMR data[27]. Compared with ileum, jejunum contained more lipids, GPC, choline, lactate and creatinine, but lower levels of amino acids and acetate[27]. Furthermore, the biochemical composition of intestinal tissues from young rats (15, 36 days, 3-4 months old) was age dependent; the levels of lipids, lactate, taurine and creatinine were positively correlated with age while amino acids and GPC decreased in the older age group[27]. Such studies were vital in providing the foundation of biochemistry for understandings to the effects lifestyle and dietary intervention on the health of the human population in modern society where stress and dietary effects are hot-topics.

2.3.2 The stress effects on rats' endogenous metabolism and LCPUFA intervention. In a recent study, the NMR-based metabonomic approach has been applied[28] to evaluate the systemic metabolic consequences of rats in response to exposure to a sole stress in the form of maternal separation, water avoidance, the combination of maternal separation followed with water avoidance later in life, and the effects of intervention with a high PUFA diet on stressed rats. Compared with the control rats, the blood plasma of mature rats exposed to the maternal separation and water avoidance stress have shown[28] similar effects with clearly decreased levels of total lipoproteins and increased levels of ketone bodies, amino acids, glucose, lactate, creatine and citrate. Acute water avoidance stress also led[28] to elevated levels of O-acetyl glycoproteins and choline in blood plasma. What is interesting is that[28] the rats exposed water avoidance alone appeared to be more separated from the controls than those exposed to the combined stressors; the early life stress seemed desensitized the rats to later life stressors. Dietary enrichment with LCPUFAs altered the lipoprotein profiles of the stressed rats but failed to reverse the stress-induced changes[28].

2.3.3 The systemic metabolic consequences of chamomile intake[29]. Natural products have been employed worldwide as alternative medicines or functional food for centuries. The flower of chamomile is marketed in UK as a common herbal beverage and is believed to possess anti-oxidative, antimicrobial, anti-inflammatory, mild sedative and antiulcer functions. These have been backed up by the fair amounts of characterization and assessment of the biological activity of individual extractable components. For example, chamomile flavonoids, such as apigenin, reduce γ-aminobutyric acid-activated chloride currents and reduce locomotor activity of rats. However, little has been studied on the response of human metabolic systems to the intake of chamomile tea until recently. In a ^{1}H NMR-based metabonomic study to understand the global metabolic response of man to chamomile tea ingestion, it has been shown[29] that (1) the effects of many confounding factors including physiological variation, daily diets and medication are in the same order of magnitude to the chamomile effects; they have to be carefully considered and disentangled from the effects of chamomile ingestion[29]; (2) the effects of chamomiles are generally weak ones, which reduced the oxidative stress indicated by the depleting levels of creatinine and affected the gut microflora metabolisms[29] indicated by the elevation of the levels of hippurate and phenylacetylglutamine. It was also striking to note that the effects of chamomile ingestion on human systemic metabolic processes were prolonged and not recovered within two weeks post-dosing period, implying a persistent disruption of the resident gut micro flora activities. It is not known whether such effects are beneficial or not and whether an eventual recovery is fulfilled. It was also unknown whether the effects were simply the sum of individual components or there were synergistic effects from its global composition, and what contributions each compositional groups made.

2.3.4 The soy effects on human endogenous metabolism[30,31]. A soy-intervention human metabonomic study showed that the high degree of inter-subject and temporal variability were present, obscuring more subtle metabolic changes in human plasma from the dietary intervention. The soy effects were subject-dependent but the nature of the response was consistent across subjects; soy intervention led to elevated levels of 3-D-hydroxybutyrate, N-acetyl glycoprotiens, and lactate together with a decrease in plasma sugar concentrations. The soy diet exerted a consistent effect on the lipoprotein profile leading to the decrease of choline level and increase in the glycerol moiety and lipoprotein groups. These observations seemed indicative that the soy intervention affected carbohydrate and energy metabolism[30]. Urinary metabonomic analysis showed that[31] soy isoflavone ingestion had significant effects on several metabolic pathways related to osmolyte fluctuation and energy metabolism. These biochemical changes were more prominent following ingestion of the unconjugated soy isoflavone (miso) diet, suggesting the presence of structure-activity relationship for the isoflavones' biological efficacy *in vivo*[31].

2.3.5 Metabolic responses of rats to polyphenol intake[32]. Although for many years polyphenols such as flavonoids consumption has been attributed to a number of potential health benefits including cancer prevention, anti-inflammatory action, and cardioprotectant activity, the biochemical effects of polyphenol consumption on living systems are not clearly understood. In an excellent example, the metabolic responses of rats to consumption of epicatechin, a polyphenolic component of tea and many common plant-originated foods, were studied by urinary metabonomic analysis. The results showed that[32], 8 hours after a single dose of 22 mg of epicatechin, there was an increase of hippurate accompanied with a decrease in urinary concentrations of succinate, citrate, 2-oxoglutarate, acetate, dimethylamine, taurine and creatinine. This indicates some interruption to TCA cycles and changes in kidney osmolyte concentrations. The decreased creatinine level and increased levels of hippurate have also been observed in studies of the chamomile[29] effects on human. The creatinine changes can be attributed to reduced oxidative stress which is not surprising given the polyphenolic nature of epicatechin whereas the changes of hippurate are associated with gut microflora. Furthermore, metabonomic approach has revealed that such effects were reversible during the 1-2 days period post-dose period. What is not clear, however, includes (1) whether epicatechin itself or its metabolites induced such urinary metabonomic changes; (2) what the effects of prolonged epicatechin consumption are on mammalian and indeed human metabolic processes; (3) whether the effects of the prolonged epicatechin consumption will also be reversible or not.

3 FUTURE PERSPECTIVES

In every studies discussed above, more questions were raised than answered and some important questions related to possible synergy were also under the spot light. This is probably the results of hypothesis generating nature of metabonomics and our limited knowledge. Nevertheless, it is clear that metabonomics is potentially a powerful tool in food quality controls and to understand the effects of food interventions on human global systems biology especially when coupled with carefully designed experimental protocols to dissect the information intermingled with other factors. The further development and applications of metabonomics in nutritional biochemistry will further develop the nutrimetabonomics and perhaps gradually bridge two distinct areas of research, namely, the efficacy of pharmaceuticals and nutriceuticals. For this, the phytometbonomics will be particularly relevant. With the exciting era of nutritional metabonomics dawning, it is expected that some more exciting applications of metabonomics in food science will be witnessed in the near future.

Acknowledgement: Tang acknowledges the financial supports from Wuhan Institute of Physics and Mathematics, the Chinese Academy of Sciences (100T Program, 2005[35]), National Natural Science Foundation of China (Grant NO.20575074). Wang acknowledges the Nestec S.A., Switzerland, for a Research Fellowship. We also acknowledge Professors Jeremy Nicholson of Imperial College London and Liping Zhao of Shanghai Jiaotong University for numerous stimulating discussions during last months.

REFERENCES

1. H. R. Tang and Y. L. Wang, *Prog. Biophys. Biochem.*, 2006, **33**, 401.
2. J. K. Nicholson, J. C. Lindon and E. Holmes, *Xenobiotica*, 1999, **29**, 1181.
3. J. K. Nicholson and I. D. Wilson, *Nat. Rev. Drug Discov.*, 2003, **2**, 668.
4. W. B. Dunn and D. I. Ellis, *TRAC-Trends Anal. Chem.*, 2005, **24**, 285.
5. O. Fiehn, *Plant Mol. Biol.*, 2002, **48**, 155.
6. M. Spraul, A. S. Freund, R. E. Nast, R. S. Withers, W. E. Maas and O. Corcoran, *Ana. Chem.*, 2003, **75**, 1536.
7. P. S. Belton, I. Delgadillo, E. Holmes, A. Nicholls, J. K. Nicholson and M. Spraul, *J. Agri. Food Chem.*, 1996, **44**, 1483.
8. P. S. Belton, I. J. Colquhoun, E. K. Kemsley, I. Delgadillo, P. Roma, M. J. Dennis, M. Sharman, E. Holmes, J. K. Nicholson and M. Spraul, *Food Chem.*, 1998, **61**, 207.
9. P. S. Belton, I. Delgadillo, A. M. Gil, P. Roma, F. Casuscelli, I. J. Colquhoun, M. J. Dennis and M. Spraul, *Magn. Reson. Chem.*, 1997, **35**, S52.
10. G. Le Gall, M. Puaud and I. J. Colquhoun, *J. Agri. Food Chem.*, 2001, **49**, 580.
11. A. M. Gil, I. F. Duarte, I. Delgadillo, I. J. Colquhoun, F. Casuscelli, E. Humpfer and M. Spraul, *J. Agri. Food Chem.*, 2000, **48**, 1524.
12. A. P. Sobolev, A. Segre and R. Lamanna, *Magn. Reson. Chem.*, 2003, **41**, 237.

13. G. Le Gall, I. J. Colquhoun, A. L. Davis, G. J. Collins and M. E. Verhoeyen, *J. Agri. Food Chem.*, 2003, **51**, 2447.
14. G. Le Gall, I. J. Colquhoun and M. Defernez, *J. Agri. Food Chem.*, 2004, **52**, 692.
15. Y. L. Wang, H. R. Tang, J. K. Nicholson, P. J. Hylands, J. Sampson, I. Whitcombe, C. G. Stewart, S. Caiger, I. Oru and E. Holmes, *Plant. Med.*, 2004, **70**, 250.
16. I. Duarte, A. Barros, P. S. Belton, R. Righelato, M. Spraul, E. Humpfer and A. M. Gil, *J. Agri. Food Chem.*, 2002, **50**, 2475.
17. I. Savary, E. Debrias, D. Dardevet, C. Sornet, P. Capitan, J. Prugnaud, P. P. Mirand and J. Grizard, *Br. J. Nutr.*, 1998, **79**, 297.
18. I. Cassar-Malek, A. Listrat and B. Picard, *Prod. Anim.*, 1998, **11**, 365.
19. A. R. Morrison and G. Mindel, *Adv. Chronic Kidney Dis.*, 2004, **11**, 197.
20. G. N. Neigh, S. L. Bowers, L. M. Pyter, M. L. Gatien and R. Nelson, *J. Endocrinology*, 2004, **145**, 4309.
21. D. Fanchimont, *Ann. N. Y. Acad. Sci.*, 2004, **1024**, 124.
22. S. V. Coutinho, P. M. Plotsky, M. Sablad, J. C. Miller, H. Zhou, A. I. Bayati, J. A. McRoberts and E. A. Mayer, *Am. J. Physiol. -Gastr. L*, 2002, **282**, G307.
23. D. Rousseau, D. Moreau, D. Raederstorff, J. P. Sergiel, H. Rupp and A. Grynberg, *Mol. Cell Biochem.*, 1998, **178**, 353.
24. L. Demaison and D. Moreau, *Cell Mol. Life Sci.*, 2002, **59**, 463.
25. C. R. Harper and T. A. Jacobson, *Arch. Intern. Med.*, 2001, **161**, 2185.
26. S. Kalmijn, M. P. J. van Boxtel, M. Ocke, W. M. M. Verschuren, D. Kromhout and L. J. Launer, *Neurology*, 2004, **62**, 275.
27. Y. L. Wang, H. R. Tang, E. Holmes, J. C. Lindon, M. E. Turini, N. Sprenger, G. Bergonzelli, L. B. Fay, S. Kochhar and J. K. Nicholson, *J. Proteome Res.*, 2005, **4**, 1324.
28. Y. L. Wang, E. Holmes, H. R. Tang, J. C. Lindon, N. Sprenger, M. E. Turini, G. Bergonzelli, L. B. Fay, S. Kochhar and J. K. Nicholson, *J. Proteome Res.*, 2006, **5**, 1535.
29. Y. L. Wang, H. R. Tang, J. K. Nicholson, P. J. Hylands, J. Sampson and E. Holmes, *J. Agri. Food Chem.*, 2005, **53**, 191.
30. K. S. Solanky, N. J. C. Bailey, B. M. Beckwith-Hall, A. Davis, S. Bingham, E. Holmes, J. K. Nicholson and A. Cassidy, *Anal. Biochem.*, 2003, **323**, 197.
31. K. S. Solanky, N. J. C. Bailey, B. M. Beckwith-Hall, S. Bingham, A. Davis, E. Holmes, J. K. Nicholson and A. Cassidy, *J. Nutr. Biochem.*, 2005, **16**, 236.
32. K. S. Solanky, N. J. C. Bailey, E. Holmes, J. C. Lindon, A. L. Davis, T. P. J. Mulder, J. P. M. Van Duynhoven and J. K. Nicholson, *J. Agri. Food Chem.*, 2003, **51**, 4139.

METABOLOMICS IN FOOD SCIENCE: EVALUATING THE IMPACT OF FUNCTIONAL FOODS ON THE CONSUMER

C.A. Daykin[1], F. Wülfert[2], J.P.M. van Duynhoven[3]

[1] Division of Molecular and Cellular Science, School of Pharmacy, University of Nottingham, University Park, Nottingham, UK, NG7 2RD.
[2] Division of Food Science, School of Bioscience, University of Nottingham, Sutton Bonington Campus, Loughborough, UK, LE12 5RD
[3] Advanced Measurement and Imaging, Unilever Food and Health Research Institute, PO Box 217, 3130 AC Vlaardingen, The Netherlands

1 INTRODUCTION

Traditionally, the term 'healthcare industry' has referred to the pharmaceutical industry and hence, the cure of disease. The food and cosmetics industries, on the other hand, have both considered their customer base to be healthy individuals. However, in recent years there has been a shift in consumer interest to prevention rather than cure of disease and this has lead to the subsequent development of numerous products aimed at individuals who do not have health problems severe enough to require medical intervention, but nevertheless whose health could be improved by dietary and lifestyle alterations. Such products include cosmaceuticals (cosmetics with active ingredients), nutriceuticals and functional foods. However, the development of any product which influences the health status of individuals requires rapid, advanced analytical tools in order to establish efficacy of the products by comprehensive analysis of measurable parameters within consumers, such as biofluid metabolites.

Conventional ('target analysis') approaches to metabolite analysis where certain compounds of interest are pre-defined and studied at the exclusion of all others are laborious and often involve elaborate sample preparation and multiple procedures; each limited to the analysis of only a few compounds or a particular class of compounds. Additionally, the extensive fractionation often required for such approaches could be fraught with artefacts such as metabolite breakdown, which contribute to variable or poor recovery. With recent advances in NMR spectroscopy, many of these problems can be circumvented. Extensive metabolite profiles of complex biological samples can be obtained with minimal sample preparation. However, the very complexity of the biochemical information derived from the NMR spectral metabolite profiles necessitates the use of advanced data reduction and pattern recognition (PR) techniques for ease of data visualisation and sample classification (diagnostic) purposes. This approach, commonly referred to as metabonomics or metabolomics has previously shown enormous potential in furthering the understanding of disease processes[1-3], toxicological processes[4-6], phenotypic outcome of gene expression[7] and biomarker discovery[8,9].

1.1 What is the difference between metabonomics and metabolomics?

As the most recent addition to the 'systems biology' toolbox, the definition and even in fact, the name, metabo*omics, is less well defined than e.g. genomics or proteomics. Hence, different authors use different definitions for similar terms or conversely, the same approach is given different names, e.g. metabonomics, metabolomics and metabolic profiling.

Whilst some research groups use these terms interchangeably, others argue that the terms refer to philosophically different aims of analysis. For example, Nicholson *et al* previously defined metabonomics as: "*quantitative measurement of multiparametric response of living organisms to (patho)physiological stimuli or genetic modification*"[10] and later extended this definition with: "*an approach to understanding global metabolic regulation of an organism and its commensal and symbiotic partners*"[11]. Nicholson argues that the name 'metabonomics' incorporates the analysis of products of non-enzymatic reactions, otherwise known as metabonates, which can interact with and influence metabolite formation [11]. Groups adhering to this definition use the term metabolomics to discuss only work dealing with simple cell systems and mainly intracellular metabolite concentrations.

For the purposes of this paper, however, the authors will use the term metabolomics to encompass both of the above scenarios.

1.2 The Role of NMR Spectroscopy in Biofluid Analysis

It is now well established that high resolution NMR spectroscopy of biofluids provides useful qualitative and quantitative biochemical information relating to metabolic status and the features of NMR spectroscopy that prove useful in this type of study have been summarised in detail elsewhere[12].

Numerous studies have shown the application of NMR spectroscopic analysis to a wide range of biofluids, including amongst others, urine[4-7,13,14], blood plasma[3,15,16], blood serum[17], seminal fluid[18], spinal fluid[19] and sweat[20]. The most appropriate sample type will be to a large extent dependent upon the research problem in question. For example, in a typical toxicology study it is possible to gain information from any biofluid or tissue sample considered necessary. In the cosmetics industry, the use of novel sample types such as sweat or skin may be more appropriate. On the other hand, in the food industry, most studies are carried out to determine effects of foods in healthy human volunteers and hence, the biofluids available for analysis are restricted to those which are easy to collect from humans, namely urine, blood plasma (or serum) and possibly saliva.

All biological fluids have characteristic physiochemical properties, and these partly dictate both the types of NMR experiment that may be employed and the experimental conditions required to maximise the biochemical information from each biofluid type. For example urine has high water content but no lipids and little/no proteins and therefore standard 1- and 2-dimensional experiments with water presaturation are generally sufficient to obtain most information.

On the other hand, plasma contains a lot of latent information, not visible by application of a simple standard 1D water presaturation experiment due to the high proportion of protein and lipids in the sample[21]. Spectral editing techniques such as the Carr-Purcell-Meiboom-Gill (CPMG) spin echo sequence or J-resolved (J-Res) spectroscopy may be applied to access information relating to low MW compounds more efficiently[21]. Diffusion editing may also be desirable in order to enhance the study of large molecules a sample, such as lipoproteins, without complicating the spectra by the presence of overlapping low molecular weight metabolites[22-24]. In the case of plasma and serum, internal probe temperature can also have a major effect on the experimental results, where variation in probe temperature allows the study of temperature dependent phase transitions within lipoprotein particles[25]. A further option is to extract the sample, in order to e.g. study protein-bound low molecular weight metabolites which will be rendered NMR invisible in whole plasma due to shortened T2 relaxation times[16]. It may also be desirable to extract lipids from blood plasma in order to gain a more detailed profile than that obtained by diffusion-editing alone. However, it should be noted that whilst the method selected for sample extraction is often given little consideration, the resultant profile can vary enormously dependent on the method selected.

2 DIFFICULTIES IN ESTABLISHING HEALTH CLAIMS FOR FOOD PRODUCTS

Although metabolomics has been applied to nutritional research in human volunteers for several years, with the first reported use of metabolomics in human nutritional studies in 2001[26], in comparison with toxicology and disease biomarker identification this is the least mature of all the metabolomics application areas because from both the analytical and chemometric perspectives it is probably the most challenging. This can be attributed to a number of complexities which are either less applicable or not applicable in other areas; 1) food products are aimed at the 'normal, healthy' population, 2) the metabolic effects of foods are much more subtle than the effects of drugs, 3) response of individuals to bioactive ingredients can vary dramatically depending upon other lifestyle factors such as age, habitual diet, exercise regime etc, 4) a food product is frequently largely uncharacterised. However, of all these problems, the first is probably the most challenging. In animal studies, whether using metabolomic or conventional bioanalysis methods, the exact biological history of the experimental animals are well controlled, but in the case of clinical studies, inherent metabolic variation in healthy people is usually large and the influence of a host of external, uncontrollable factors on the metabolite profile of human samples renders the definition of normality problematic.

3 STRATEGIES TO DEAL WITH HUMAN DIVERSITY IN A NUTRITIONAL TRIAL

Strategies to deal with human diversity in nutritional trials can be divided into two categories; 1) implicit strategies and 2) explicit strategies. Implicit strategies are those which are purely design driven. In this case, factors which are likely to introduce any unnecessary diversity into the study are identified and considered at the pre-volunteer recruitment stage of the study and incorporated in the study design. Explicit strategies are those where a computational correction method is applied after spectral acquisition in order to limit the effects of any remaining unwanted variation in the data. The two strategies are not mutually exclusive and should be used in combination with one another.

3.1 Implicit Strategies

It should be remembered that everything, whether genetic or environmental will affect the metabolic profile of an individual. Factors well established to produce notable effects on the metabolic profile of individuals include; age, gender, body mass index (BMI), time of sample collection, drugs (whether prescription or non-prescription, including caffeine), habitual diet (including vegetarian, alcohol, fish, certain vegetables) and the amount of exercise regularly taken. These factors should therefore all be taken into consideration in a well designed study. However, even the best designed human trial will contain flaws.

3.1.1 Implicit strategy flaw 1 - What is 'healthy'? When designing a study, defining a healthy, control population can be problematic, after all, what is meant by 'healthy'? 'Apparently healthy' is simply the absence of diagnosed disease and it can be possible that volunteers have disease(s) of which they are unaware. Furthermore, since there are ethical implications of screening people for disease prior to pronouncing them suitable as control subjects, it will be necessary to accept that not all control, healthy individuals will be as healthy as first apparent.

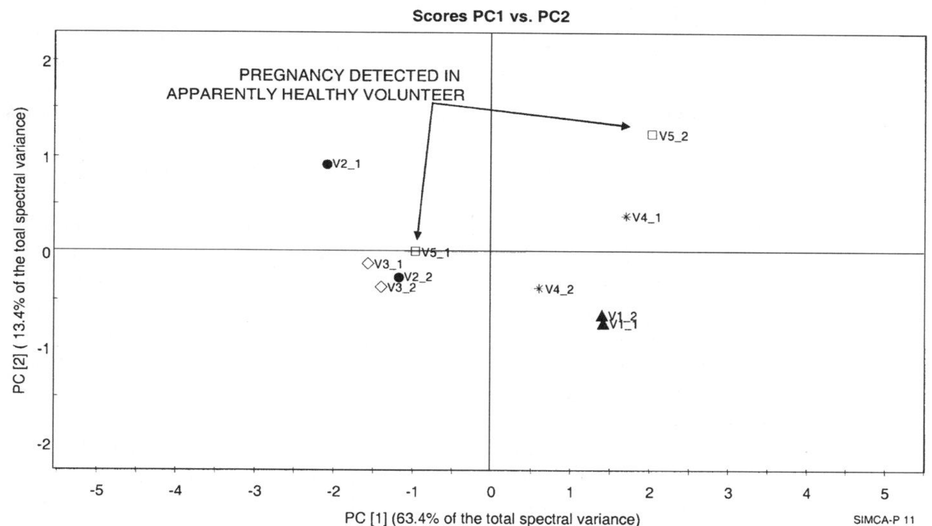

Figure 1 Principal component analysis (PCA) scores plot from a study where volunteers were requested to provide two plasma samples, a number of weeks apart. Symbols on the score plot indicate the different volunteers and it can clearly be seen that in the case of most volunteers, their two plasma samples are located very close together on the scores plot, indicating little compositional change of the blood plasma collected. In the case of volunteer 5, however (shown here as V5_1 and V5_2, representing the first and second collection times, respectively the samples are located a long way apart on the scores plot, indicating a large metabolic change in this individual, later found to be due to pregnancy of this volunteer.

3.1.2 Implicit strategy flaw 2 - Volunteer compliance with dietary restrictions. When carrying out studies into the effects of dietary changes on the metabolism, it is usually necessary to ask volunteers to remove the food or beverage under study from their diet for several days, but not all volunteers will comply with study instructions, frequently as a result of simple mistakes or forgetfulness on the part of volunteers. An example is illustrated in figure 2. This spectrum was taken from a study in which volunteers were requested to refrain from consumption of artificial sweeteners but the sweetener acesulfame K was detected in this volunteer's urine, thus showing that the volunteer had not complied with study instructions.

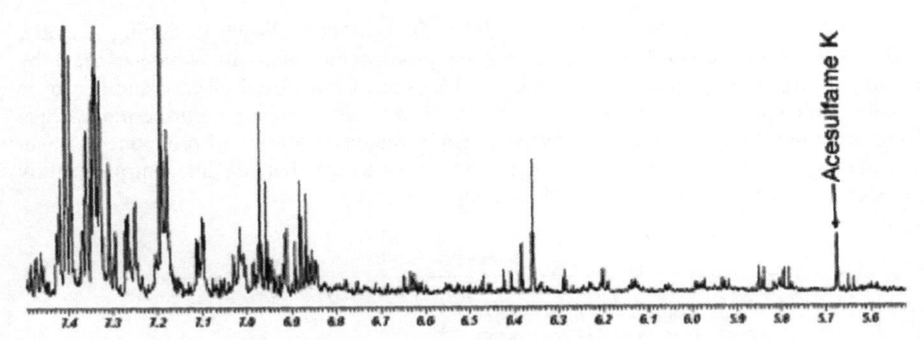

Figure 2 Partial 600 MHz ^{1}H NMR spectrum (δ 7.5-5.5) of urine collected from a male volunteer. The volunteer had consumed a sweetened cola drink, as evidenced by the presence of sweetener acesulfame K in his urine.

3.1.3 Implicit strategy flaw 3 – Imposing dietary restrictions on volunteers. In cases where the food product or ingredient of interest is frequently consumed, it may be necessary to ask volunteers to make relatively large alterations to their habitual diet prior to the study. An example of this is in the study of flavonoids. Flavonoids are abundant compounds in many different types of food and drink such as fruits, vegetables, tea, coffee, beer or red wine and hence, removal of all flavonoid-containing foods and beverages from the diet is likely to induce metabolic changes in volunteers. An example of this is shown in figure 3, which shows a number of urine spectra collected over a 5-day study period[13]. On days 1-3, volunteers were requested to follow a low-flavonoid diet and on the morning of day 4, volunteers were given strong black tea. The effects of the low flavonoid diet can be clearly observed in the urine spectra from days 1-3. In particular, attention is drawn to the hippurate peaks highlighted, which progressively decrease during the dietary restriction period. However, methods which minimise the requirement for this dietary restriction are currently under investigation by the authors.

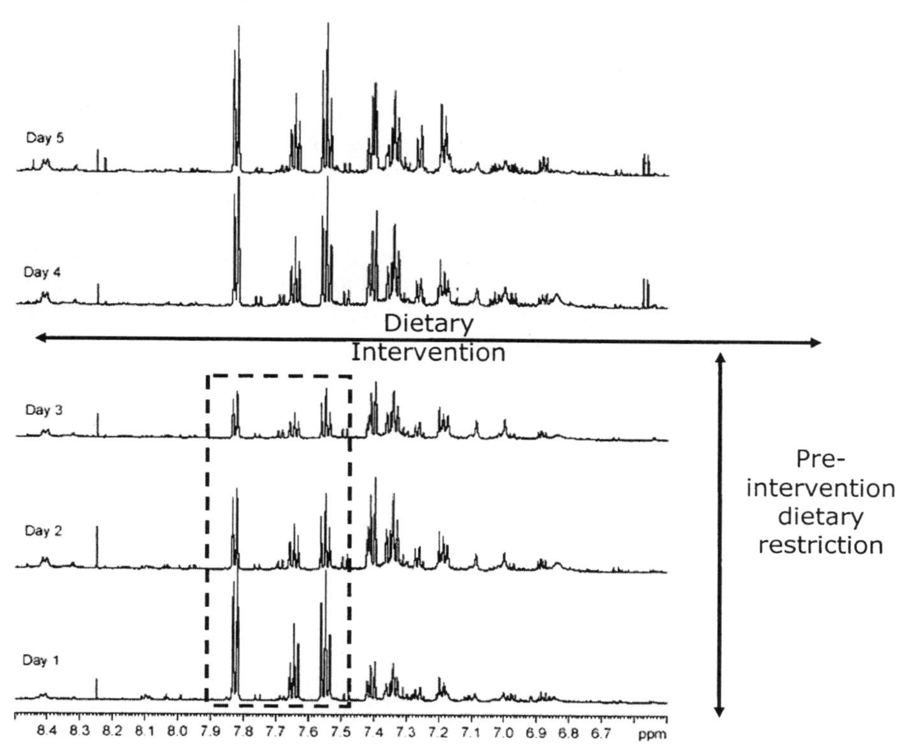

Figure 3 The aromatic region of ^{1}H NMR spectra (δ 8.5-6.5) of 24 hour urine samples from a volunteer following a five day study where the volunteer was requested to follow a low flavonoid diet from days 1-3, prior to being given a single dose of strong black tea on the morning of day 4

3.2 Explicit strategies

There are a multitude of different options for computational removal of unwanted variation within a metabolomic data set and again, the most appropriate option will be dependent upon the research question. In a typical disease biomarker discovery project, the application of a predicting or 'supervised' data analysis method i.e. discriminant analysis will frequently be sufficient to exclude any interfering factors from the data analysis model. However, in most nutritional studies, this will be insufficient. One option which is useful in longitudinal (time-course) studies, is to subtracting for each person his or her 'typical personal spectrum'. A good estimator for this personal spectrum is the average spectrum of each person, this will then allow each individual's variation around their own average state to be studied. A similar approach can also be applied where day-to-day variation in the spectrometer operation causes confounding factors, provided that spectrometer variation can be separated from group or individual variation. Therefore, the importance of acquiring data in a random order or considering sample order in the study design can not be overstated. Another option is the subtraction of spectra from biofluids collected prior to dietary intervention i.e. those taken at the beginning of the study, in order to focus only on the study-induced metabolic changes. Different scaling strategies can then be applied to further reduce the person dependent effects.

4 CONCLUSIONS

In conclusion, it can be seen that the design of nutritional trials is far from straight forward, largely due to the fact that the metabolic effects under investigation are much smaller than the inherent metabolic diversity of volunteers. Thus, in the case of nutritional studies, a good study design which minimises volunteer diversity is essential. However, this will not usually be sufficient and it will therefore be necessary to apply computational correction methods to further reduce sources of metabolic variation which were impossible to remove through study design. After implementing these steps, metabolomics can even be used to investigate even very subtle metabolic influences. This is essential in a changing health care market, where interests increasingly focus on prevention rather than cure of disease.

References

1 J.K. Nicholson, M. O'Flynn, P.J. Sadler, A. Macleod, S.M. Juul, P.H. Sönksen, *Biochem. J.*, 1984, **217,** 365–375.
2 A.M. Tomlins, P.J.D. Foxall, J.C. Lindon, M.J. Lynch, M. Spraul, J.R. Everett, J.K. Nicholson, *Anal. Comm.*, 1998, **35**, 113-115
3 J.T. Brindle, H. Antti, E. Holmes, G. Tranter, J.K. Nicholson, H.W.L. Bethell, S. Clarke, P.M. Schofield, E. McKilligin, D.E. Mosedale, D.J. Grainger, *Nature Medicine*, 2002, **8**, 1439-1444
4 K.P.R. Gartland, S.M. Sanins, J.K. Nicholson, B.C. Sweatman, C.R. Beddell, J.C. Lindon, *NMR Biomed.*, 1990, **3**, 166-172.
5 K.P.R. Gartland, C.R. Beddell, J.C. Lindon, J.K. Nicholson, J.K., *Mol. Pharmacol.* 1991, **39**, 629-642.
6 M.L. Anthony, C.R. Beddell, J.C. Lindon, J.K. Nicholson, *J. Pharm. Biomed. Anal.*, 1993, **11**, 897-902
7 C.L. Gavaghan, E. Holmes, E.M. Lenz, I.D. Wilson, J.K. Nicholson, *FEBS Lett.*, 2000, **484**, 169-174
8 P.J.D. Foxall, S. Bewley, G.H. Neild, C.H. Rodeck, J.K. Nicholson, J.K., *Arch. Dis. Child Fetal Neonatal Ed.*, 1995, **73**, F153-7.
9 J.J. Powell, K.P.R.Gartland, M. Lombard, R. Sallie, J.K. Nicholson, R.P.H. Thompson, *Clin. Sci.* 1990, **78**, 13.
10 J.K. Nicholson, J.C. Lindon, E. Holmes, *Xenobiotica*, 1999, **11**, 1181-1189.
11 J.K. Nicholson, I.D. Wilson, *Nature Drug Discovery*, 2003, **2**, 668-677.
12 C.A. Daykin, F. Wülfert, *Frontiers in Drug Design and Discovery*, 2006, **2**, 151-173,
13 C.A. Daykin, J.P.M. van Duynhoven, A. Groenewegen, M.Dachtler, J.M.M. van Amelsvoort, T.P.J. Mulder, *J. Food Agric. Chem.*, 2005, **53**, 1428-1434
14 F.A. van Dorsten, C.A. Daykin, T.P.J. Mulder, J.P.M. van Duynhoven, *J. Food Agric. Chem.*, 2006, In Press
15 R.A. Wevers, U. Engelke, A. Heerschap, *Clin. Chem.*, 1994, **40**, 1245-1250.

16 C.A. Daykin, P.J.D. Foxall, S.C. Connor, J.C. Lindon, J.K. Nicholson, *Anal. Biochem.*, 2002, **304**, 220-230
17 C.A. Daykin, O. Corcoran, S.H. Hansen, I. Bjørnsdottir, C. Cornett, S.C. Connor, J.C. Lindon, J.K. Nicholson, *Anal. Chem.*, 2001, **73**, 1084-1090
18 A.M. Tomlins, P.J.D. Foxall, M.J. Lynch, J.A. Parkinson, J.R. Everett, J.K. Nicholson, *Biochim. Biophys. Acta*, 1998, **1379**, 367-380
19 S. Maillet, J. Vion-Dury, S. Confort-Gouny, F. Nicoli, N.W. Lutz, P. Viout, P.J. Cozzone, *Brain Research Protocols*, 1998, **3**, 123-134.
20 M. Harker, H. Coulson, I. Fairweather, D. Taylor, C.A. Daykin, Metabolomics, 2006, In Press (available online at http://dx.doi.org/10.1007/s11306-006-0024-4)
21 J.K. Nicholson, P.J.D. Foxall, M. Spraul, R.D. Farrant, J.C. Lindon, *Anal. Chem.*, 1995, **67**, 793-811.
22 C.A. Daykin, *Novel NMR and HPLC-NMR Studies of Blood Plasma*, PhD Thesis, Imperial College, University of London, **2000**
23 J.L. Griffin, H.J. Williams, E. Sang, E., J.K. Nicholson, *Magn. Reson. Med.*, 2001, **46**, 249-255
24 B.M. Beckwith-Hall, N.A. Thompson, J.K. Nicholson, J.C. Lindon, E. Holmes, *Analyst*, 2003, **128**, 814-818
25 M. Ala-Korpela M, *Prog. NMR Spectrosc.,* 1995**,** **27**, 475-554
26 C.A. Daykin, J.P.M. van Duynhoven, A. Groenewegen, T.P.J. Mulder, J.M.M. van Amelsvoort, *^{1}H NMR spectroscopic studies on the metabolic impact of black tea in humans*, 42nd ENC, Florida, USA. 2001.

Food Quality

^{1}H NMR-BASED METABONOMICS APPLIED IN THE ELUCIDATION OF BIOCHEMICAL EFFECTS OF CONSUMPTION OF WHOLE GRAIN CEREALS

H.C. Bertram[1], K.E. Bach Knudsen[2], A. Malmendal[3], N.C. Nielsen[3], X. Fretté[1] and H.J. Andersen[4]

[1]Department of Food Science and [2]Department of Animal Health, Welfare and Nutrition, Danish Institute of Agricultural Sciences, Research Centre Foulum, DK-8830 Tjele, Denmark, [3]Center for Insoluble Protein Structures (inSPIN), Interdisciplinary Nanoscience Center (iNANO) and Dept. Chemistry, University of Aarhus, Denmark, [4]Arla Foods amba, Corporate R&D, Skanderborgvej 277, DK-8260 Viby J, Denmark

1 INTRODUCTION

Metabonomics is defined as "The quantitative measurement of the time-related multi-parametric metabolic response of living systems to pathophysiological stimuli or genetic modification"[1]. Several techniques exist for generating metabonomics data, however, high-resolution ^{1}H NMR spectroscopy has proven to be one of the most powerful technologies for examining and obtaining metabonomic data on biofluids[2]. A major advantage of the NMR technique is the minimal sample preparation and its non-selectivity, which implies that fluctuations in an entire range of metabolites can be detected. The disadvantage of the NMR technique is that it cannot compete with the much higher sensitivity of MS techniques. Several NMR-based metabonomic studies exploring the biochemical effects of various toxins have been reported, for a review see Lindon et al.[3]. Within the field of nutrition the use of NMR-based metabonomics is more rare, however, the technique is now also emerging quickly within this field, and applications of NMR-based metabonomic for elucidating the biochemical effects of dietary bioactive compounds in both animals[4] and humans[5-8] are demonstrated. Moreover, an NMR-based application for identification of obese-related metabolites is reported[9]. These studies have demonstrated changes in plasma lipoprotein, amino acid, and carbohydrate profiles following intervention of a certain compound. Accordingly, the NMR method provides a useful tool for non-selective, explorative investigations within the field of nutrition and for elucidating the relationship between nutrition and health.

Multiple life-style and environmental factors, including diet, influence the risk to develop common Western life-style diseases. According to WHO, 30 % of the cancers in Western societies are related to diet. Consumption of whole grain cereals has been associated with many health benefits, including a reduction of the risk for developing chronic life-style diseases such as cardiovascular diseases (CVD), diabetes, obesity and some cancers. Whole grain cereals are rich in dietary fibre, vitamins, minerals and bioactive compounds such as phytochemicals and micronutrients. Several of these compounds must be expected to target signalling pathways and biological processes relevant for the development of these diseases. Here we present an application of ^{1}H NMR-based metabonomics for elucidating the biochemical effects of whole grain rye- and wheat-based diets using the pig as a model for human subjects, which is also published elsewhere[10].

2 MATERIALS AND METHODS

2.1 Diets and animals

Two diets with similar levels of dietary fibre and macronutrients, but with contrasting levels of whole grain constituents and plant phytochemicals, were prepared from whole grain rye (WGR) and non-whole grain wheat (NWG), the exact composition of the diets is published elsewhere[10]. The experiments were divided into two sub-studies and consisted of a total of 10 pigs with an initial weight of about 45 kg. The pigs were fed the two diets in a cross-over design. Plasma and urine samples were collected after 5 and 7 days on each diet, respectively. The plasma samples were collected in heparinized plastic tubes and centrifuged (1000 x *g* in 10 min. at 8°C) to separate the red blood cells from plasma. The plasma was kept frozen at –20 °C until analysis. For urine collection the pigs were equipped with a urine bladder catheter for urine collection (5.0 mm., i.d.), and all the urine produced over 3 days was collected in benzoic acid via catheters.

2.2 NMR measurements

The NMR measurements were performed at 300 K on a Bruker Avance 400 NMR Spectrometer (Bruker BioSpin), operating at a ^1H frequency of 400.13 MHz, and equipped with a standard 5-mm HX inverse probe.
Prior to the measurements plasma samples (n=16) were thawed and 200 μl aliquots were mixed with 400 μl D_2O. Sodium trimethylsilyl-[2,2,3,3-$^2H_4$]-1-propionate (TSP) was added as an internal chemical shift standard (0.7 mg/ml). 1D ^1H NMR spectra of plasma samples were obtained using a Carr-Purcell-Meiboom-Gill (CPMG) delay added in order to attenuate broad signals from high molecular weight components. The total CPMG delay was 40 ms and the spin-echo delay was 200 μs. Urine samples (n=12) were thawed, 400 μl aliquots were mixed with 200 μl D_2O, and TSP was added as an internal chemical shift standard (0.7 mg/ml). 1D ^1H NMR spectra of urine samples were obtained using a single 90° pulse experiment. Both on plasma and urine samples water suppression was achieved by irradiating the water peak during the relaxation delay of 2 s. A total of 128 transients of 8 K data points spanning a spectral width of 24.03 ppm was collected. An exponential line-broadening function of 3 Hz was applied to the FID prior to Fourier transform. All spectra were referenced to the TSP signal at 0 ppm.
NMR measurements on plasma samples were also performed upon spiking of plasma samples with betaine (Sigma Chemical Co., St. Louis MO), glycerophoshorylcholine (Sigma Chemical Co.), enterolactone (Labmaster Ltd.), and alkylresorcinol C15:0.

2.3 Post-processing of NMR data

The mean-normalized ^1H NMR spectra in the region 9.0-5.1 ppm and the region 4.6-0.5 ppm were used for further data analysis. The spectra were segmented into regions of 0.04 ppm width, and the integral of each region was calculated. The reduced spectra consisting of 195 integrated regions were normalised to the whole spectrum to remove any concentration effects and used for further analysis, which was performed using the Unscrambler software version 9.2 (Camo, Oslo, Norway). Principal component analysis

(PCA) was applied to the centered data to explore any clustering behaviour of the samples, and partial least square regression discriminant analysis (PLS-DA) was performed to explore intrinsic biochemical dissimilarities between predefined sample classes (WGR diet vs. NWG diet). During all regressions, Martens' uncertainty test[11] was used to eliminate noisy variables, and all models were validated using full cross-validation[12].

3 RESULTS AND DISCUSSION

Typical one-dimensional ^{1}H NMR spectra acquired on blood plasma and urine are shown in Figure 1a-b. On blood plasma standard ^{1}H spectra include resonances from high-molecular weight molecules, and show principally broad resonances arising from the high amount of fatty acyl chains in lipid molecules present in the plasma. The broad resonances partially obscure the sharper peaks from low-molecular weight compounds, and therefore a CPMG ^{1}H NMR experiment was applied on blood plasma (Figure 1a). In urine the amount of high-molecular weight molecules are much less, and a standard ^{1}H experiment gives relatively well-resolved spectra (Figure 1b).

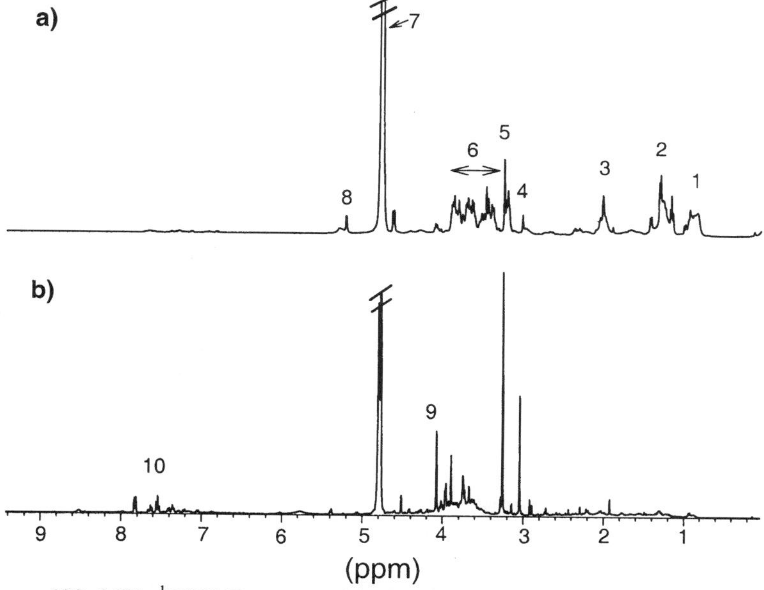

Figure 1. 400 -MHz ^{1}H NMR spectra. (a) 1D spin-echo CPMG spectrum acquired on blood plasma; (b) standard 1D spectrum acquired on urine. Assignment: 1, lipid CH_3; 2, lipid $(CH_2)_n$; 3, lipid $CH_2CH{=}CH$; 4, creatine/creatinine CH_3; 5, $N(CH_3)_3$ in choline, phoshocholine, glycerophosphocholine, TMAO and betaine; 6, CH in sugars and amino acids; 7, residual HDO; 8, C1H in α-glucose; 9, creatine/creatinine CH_2, 10, aromatic region (hippurate, urea).

In order to identify possible minor metabolic differences between sample classes, principal component analyses (PCA), which is an unsupervised method, was performed on the obtained ^{1}H spectra. For both blood plasma and urine the scatter plot shows a clear tendency for clustering according to diet, however the effect is most pronounced in urine (Figure 2a-b). This finding discloses a significant effect on the metabolite plasma and urine profile of the two diets.

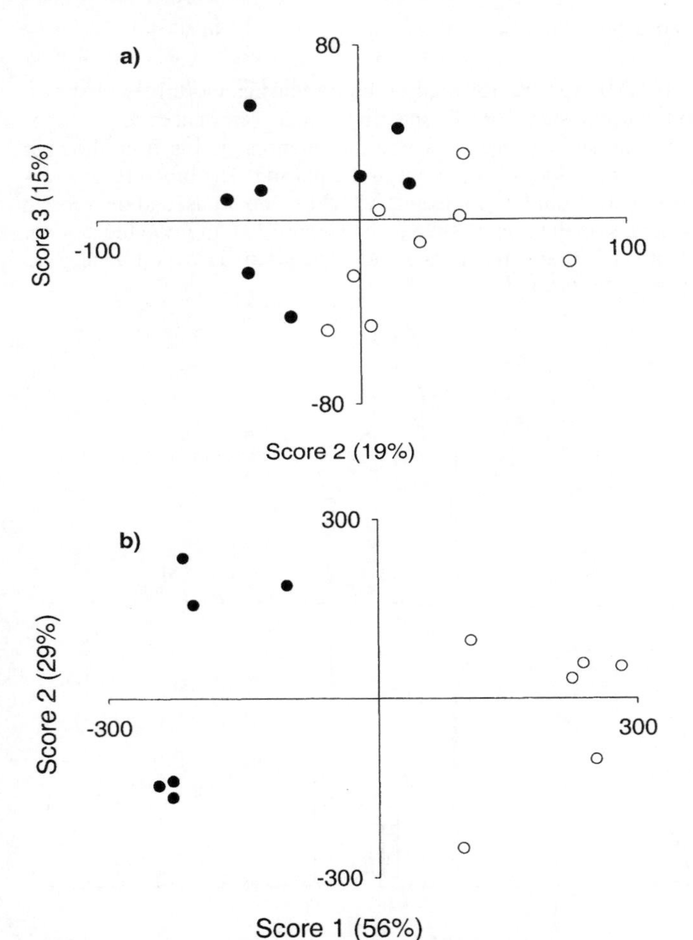

Figure 2. PCA analysis. (a) Score plot showing the second and third principal component for PCA carried out on ^{1}H CPMG spectra obtained on blood plasma samples; (b) Score plot showing the first and second principal component for PCA carried out on ^{1}H standard spectra obtained on urine samples. Filled symbols: Non-whole grain wheat (NWG) samples, open symbols: Whole grain-rye (WGR) samples.

In order to elucidate the spectral differences discriminating between the two diets, partial least squares regression discriminant analysis (PLS-DA), which is a supervised method focusing on discriminating between pre-defined classes, was performed on the plasma ^{1}H

NMR spectra to investigate the metabolic differences in plasma profile between the two diets. A clear separation of two diets was observed in the PLS-DA scores plot (data not shown). Jack-knifing revealed that the regions of the NMR spectrum which most strongly influence separation between NWG and WGR plasma are the 3.54 ppm, 3.25 ppm and 1.28 ppm regions. In order to identify the 3.25 ppm signal, blood plasma was spiked with various compounds, and ^{1}H spectra were acquired. Figure 3a-b shows spectra acquired on samples spiked with betaine and glycerophosphorylcholine. It is noteworthy that betaine gives resonances at two positions, where resonances are also present in unspiked plasma (Figure 3a), while glycerophosphorylcholine gives gives rise to three signals where only one is present in unspiked plasma. These spiking experiments therefore indicated that the 3.25 ppm and 3.54 ppm signals should be ascribed to betaine.

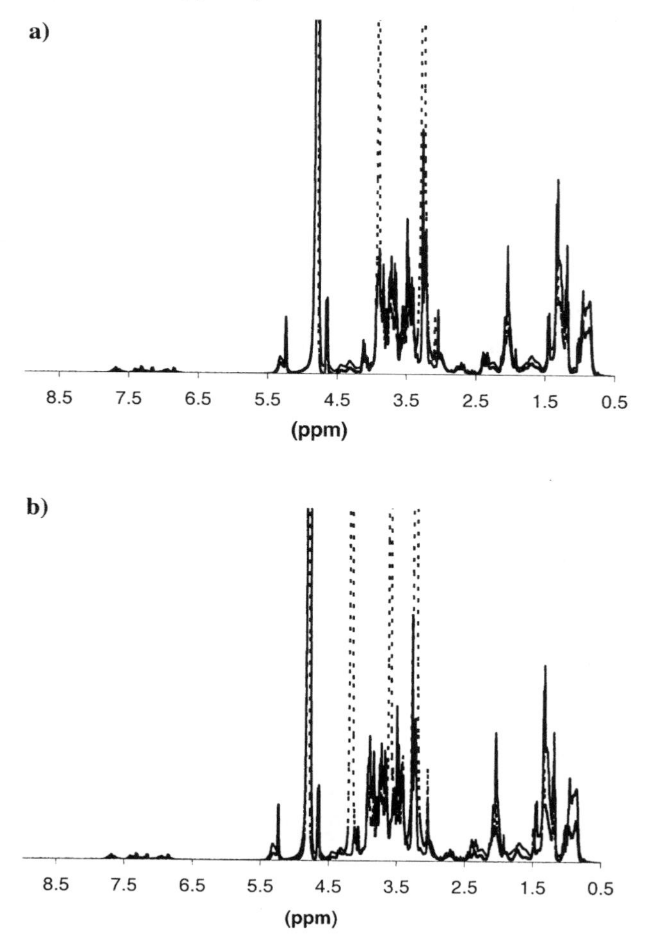

Figure 3. ^{1}H spectra of an representative plasma sample before (full line) and after (dotted line) spiking with (a) betaine and (b) glycerophosphylcholine.

Further evidence that the 3.25 ppm signal was reflecting a higher content of betaine in WGR plasma compared with NWG plasma was obtained by LC-MS analyses, which revealed that the betaine content was significantly higher in WGR plasma compared with NWG plasma (Figure 4).

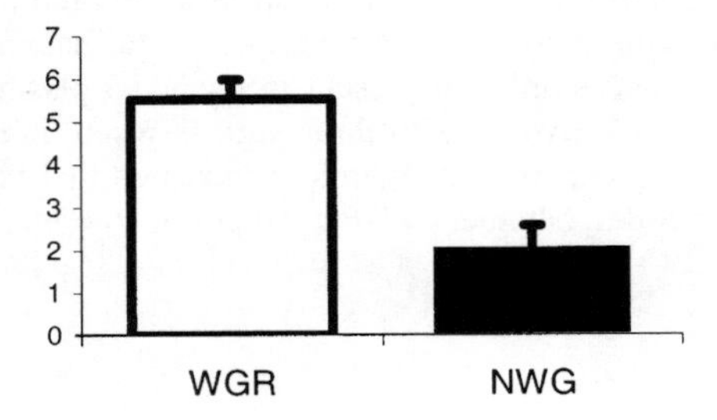

Figure 4. Betaine content (µg/ml) in whole grain rye (WGR) and non-whole grain wheat (NWG) plasma samples determined by LC-MS analysis. LSMean values are given. Bars show standard errors (n=8).

The 1.28 ppm signal is reflecting fatty acyl methylene protons. The fatty acids are partly bound in HDL and LDL lipoproteins fractions[13], and the observed effect of diet on the 1.28 ppm signal is most probably reflecting the diet-induced alteration in the HDL/LDL-fraction, which can be expected to contribute to the health benefits associated with consumption of whole grain rye cereals. Accordingly, this study demonstrates that information on changes in lipoprotein profile can be obtained using the present technique. PLS-DA was carried out on the ^{1}H spectra obtained on urine. Also in urine the spectral region at 3.25 ppm dominates the differentiation between the two diets. However, other regions showing pronounced differences between WGR and NWG urine were also identified, these are summarized in Table 1.

Table 1. Summary of the major differences between WGR and NWG urine samples

NMR spectral region	Assignment*
3.03 ppm	N-CH_3 in creatinine/creatine
3.25 ppm	N-CH_3 in betaine
3.90-3.94 ppm	CH_2 in betaine
4.06-4.10 ppm	N-CH_2 in creatinine
5.69-5.81 ppm	NH_2 in urea
7.54 ppm	CH3/CH5 in hippurate
7.83 ppm	CH2/CH6 in hippurate

The concentration of creatinine in the urine samples was confirmed and quantified using LC-MS, and a significantly higher content of creatinine was found in NWG urine samples compared with WGR urine samples (Figure 5).

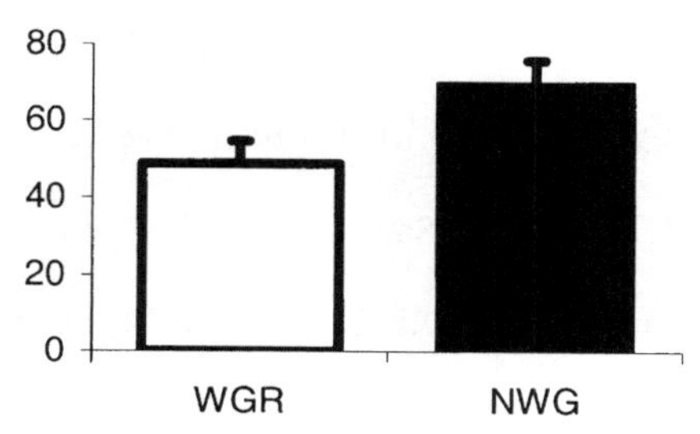

Figure 5. Creatinine content (μg/ml) in whole grain rye (WGR) and non-whole grain wheat (NWG) urine samples determined by LC-MS analysis. LSMean values are given. Bars show standard errors (n=12).

4 CONCLUSIONS

Using NMR-based metabonomics as an explorative approach in an intervention study on pigs, the present studies disclosed metabolic effects of a whole grain diet on the content of betaine in plasma and excretion of betaine and creatinine, which may contribute to the health benefits of a high dietary intake of whole grain. Further studies emphasising the beneficial role of betaine and its potential connection with creatinine excretion in the health-promoting effect of wholegrain cereals are needed. In conclusion, the present study demonstrated that ^{1}H NMR-based metabonomics on biofluids can be successfully applied for investigating the biochemical effects of diet components and thereby the techniques provides a powerful tool for obtaining a better understanding of the relationship between nutrition and health. Future studies combining ^{1}H NMR-based metabonomics on bio fluids with ^{1}H high-resolution magic angle spinning NMR studies on tissues, thereby approaching 'whole systems biology', may strengthen the technique even further.

Acknowledgements
We thank the Nordic Industrial Fund, the Danish Agricultural and Veterinary Research Council, the Danish Technology and Production Research Council (project no. 274-05-339: "NMR-based metabonomics on tissues and biofluids"), Cerealia, Sweden; Wasabröd, Sweden; Vaasan & Vaasan, Finland, Fazer Oululainen, Finland, the Danish Biotechnological Instrument Centre, Carlsbergfondet, and the Danish National Research Foundation for suppport.

References
1 J.K. Nicholson, J. Connelly, J.C. Lindon and E. Holmes, *Nat. Rev. Drug Discovery*, 2002, **1**, 153.
2 J.C. Lindon, E. Holmes and N.K. Nicholson, *Anal. Chem.*, 2003, **75**, 385.
3 J.C. Lindon, E. Holmes and N.K. Nicholson, *Progr. Nucl. Magn. Reson. Spectr.*, 2004, **45**, 109.
4 K.S. Solanky, N.J.C. Bailey, E. Holmes, J.C. Lindon, A.L. Davis, T.P.J. Mulder, Van Duynhoven, J.P.M. and J. K. Nicholson, *J. Agric. Food Chem*, 2003, **51**, 4139.
5 K.S. Solanky, N.J.C. Bailey, B.M. Beckwith-Hall, A. Davis, S. Bingham, E. Holmes, J.K. Nicholson and A. Cassidy, *Anal. Biochem.*, 2003, **323**, 197.

6 K.S. Solanky, N.J.C. Bailey, B.M. Beckwith-Hall, A. Davis, S. Bingham, E. Holmes, J.K. Nicholson and A. Cassidy, *J. Nutr. Biochem*, 2005, **16**, 236.
7 C.A. Daykin, J.P.M Van Duynhoven, A. Groenewegen, M. Dachtler, J.M.M. Van Amelsvoort and T.P.J. Mulder, *J. Agric. Food Chem*, 2005, **53**, 1428.
8 Y.L. Wang, H.R. Tang, J.K. Nicholson, P.J. Hylands, J. Sampson and E. Holmes, *J. Agric. Food Chem*, 2005, **53**, 191.
9 N.J. Serkova, M. Jackman, J.L. Brown, T. Liu, R. Hirose, J.P. Roberts, J.J. Maher and C.U. Niemann, *J. Hepatology*, 2006, **44**, 956.
10 H.C. Bertram, K.E. Bach Knudsen, A. Serena, A. Malmendal, N.C. Nielsen, X.C. Fretté and H.J. Andersen, *Br. J. Nutr.*, 2006, **95**, 955.
11 H. Martens and P. Dardenne *Chemometrics Intell. Lab. Syst.*, 1998, **44**, 99.
12 H. Martens and M. Martens, *Food Qual. Preference*, 2000, **11**, 5.
13 C.A. Daykin, O. Corcoran, S.H.Hansen, I. Bjørnsdottir, C. Cornett, S.C. Connor, J.C. Lindon and J.K Nicholson, *Anal. Chem.*, 2001, **73**, 1084.

LOW MOLECULAR WEIGHT METABOLITES IN WHITE MUSCLE FROM COD (*GADUS MORHUA*) AND HADDOCK (*MELANOGRAMMUS AEGLEFINUS*) ANALYZED BY HIGH RESOLUTION ^{1}H NMR SPECTROSCOPY

I. B. Standal[1, 2], I. S. Gribbestad[3], T. F. Bathen[3], M. Aursand[1] and I. Martinez[1].

[1]SINTEF Fisheries and Aquaculture Ltd., N-7465 Trondheim, Norway;
[2]Department of Biotechnology, Norwegian University of Science and Technology (NTNU), N-7491 Trondheim, Norway;
[3]Department of Circulation and Medical Imaging, Faculty of Medicine, Norwegian University of Science and Technology (NTNU), N-7489 Trondheim, Norway

1 INTRODUCTION

Fish is considered as healthy food because of its content of n-3 polyunsaturated fatty acids (PUFAs), proteins and small bioactive compounds such as taurine, anserine, betaine and trimethylamine N-oxide (TMAO). PUFAs play a preventive role in cardiovascular diseases and in the alleviation of other health problems. Regarding the function of the small molecules, some of them stabilize protein structure[1] or regulate osmotic pressure in cells[2], while others have additional functions; taurine is an essential growth factor, and necessary for regulating the function of eyes, heart, muscles, brain and central nervous system[3,4,5], anserine is an antioxidant[6] and betaine is important for a proper liver function[7].

Previous studies used ^{1}H NMR spectroscopy to study changes in cod (*Gadus morhua*) and haddock (*Melanogrammus aeglefinus*) muscle during frozen storage[8]. Sitter *et al.* applied ^{1}H NMR to monitor several metabolites simultaneously during frozen storage of Atlantic halibut (*Hippoglossus hippoglossus L.)*[9]. Gribbestad *et al.* gave a detailed assignment of the ^{1}H NMR spectra from Atlantic salmon[10], while Martinez *et al.* applied the technique to evaluate changes in bioactive components following freezing, thawing, cooking and salting of cod[11].

The present study reports for the first time a full interpretation of the ^{1}H NMR spectrum of perchloric extracts from cod (*Gadus morhua*) and haddock (*Melanogrammus aeglefinus*), in order to identify compounds reflecting the physiological state and nutritional value of fish.

2 MATERIALS AND METHODS

2.1 Fish

Cod (*Gadus morhua*) (n=5, 3.6+0.6 kg) and haddock (*Melanogrammus aeglefinus*) (n=5, 0.34+0.05 kg) were caught in the Trondheims fjord (March and April respectively) and held live in a big landing net for about 6 h. Fish were individually taken out from the net and killed by a blow to the head within 10-20s. Samples of white muscle (1-2 g) were excised right under the first dorsal fin, accurately weighted and frozen in liquid nitrogen within 30-60 s after slaughter.

Perchloric acid (PCA) extraction was performed in ice according to Glonek *et al.*[12]. Ice-cold PCA (2mL, 7% w/v) was added accurately weighted frozen fish sample (1-2 g) and homogenized to a paste consistency (10-20s) with an Ultra Turrax homogeniser. Ice-cold PCA (1 mL, 7% w/v) was added to remove the residue on the homogeniser blade. The homogenate was centrifuged (3500 × g, 10 min, 4 ° C). The supernatant was transferred to a new tube, while the extraction of the pellet was repeated twice as described above. All supernatants were pooled and neutralized (pH=7) with 2 M KOH in ice before centrifugation (3500 × g, 10 min, 4 ° C). The resulting supernatant was stored at -80°C until lyophilization.

2.2 NMR spectroscopy

The lyophilized extracts were redissolved in 0.7 mL of D2O containing trimethylsilypropionate-2,2,3,3-d4 (TSP, 0.5mM) and transferred to 5-mm NMR sample tubes. High resolution ^{1}H NMR spectra were recorded at ambient temperature on a Bruker Avance DRX500 spectrometer. The ^{1}H spectra were obtained using water presaturation in the relaxation delay. A sweep width of 10 kHz was collected into 32K points. Number of scans was set to 512. The raw data were multiplied with a 0.3 Hz exponentional line-broadening factor before Fourier transformation into 64K data points.

Some extract samples were spiked with small amount of the pure compounds (taurine, creatine, β-alanine, choline and anserine) to ensure correct assignment of peaks. The number of scans in the analyses of the pure compounds in D2O was reduced to 16.

The ^{1}H homonuclear correlated spectra (COSY) were obtained by applying water presaturation during relaxation delay and using gradient pulses for selection. The COSY spectra were recorded by acquisition of 90 transients per increment for 512 increments collected into 4K data points. A spectral width of 12 kHz was used in both dimensions. The time domain was zero-filled and apodized with a squared sine window function in both dimensions before Fourier transformation.

3 RESULTS AND DISCUSSION

3.1 Assignment of spectra

The 1D ^{1}H NMR spectra of the perchloric acid extracts from cod are given in Fig. 1. (Table of the assigned metabolites is available on request). The assignments were made using two dimensional ^{1}H homonuclear correlated spectroscopy (COSY), by spiking with some pure compounds and by comparison with ^{1}H NMR spectra of pure compounds in D2O. Additionally, the resonances of the compounds identified in the present samples were compared to those of the same compounds from other biological materials including human samples[13,14,15], plants[16], Atlantic Halibut (*Hippoglossus hippoglossus L.*)[9] and Atlantic salmon (*Salmo salar*)[10] for further confirmation.

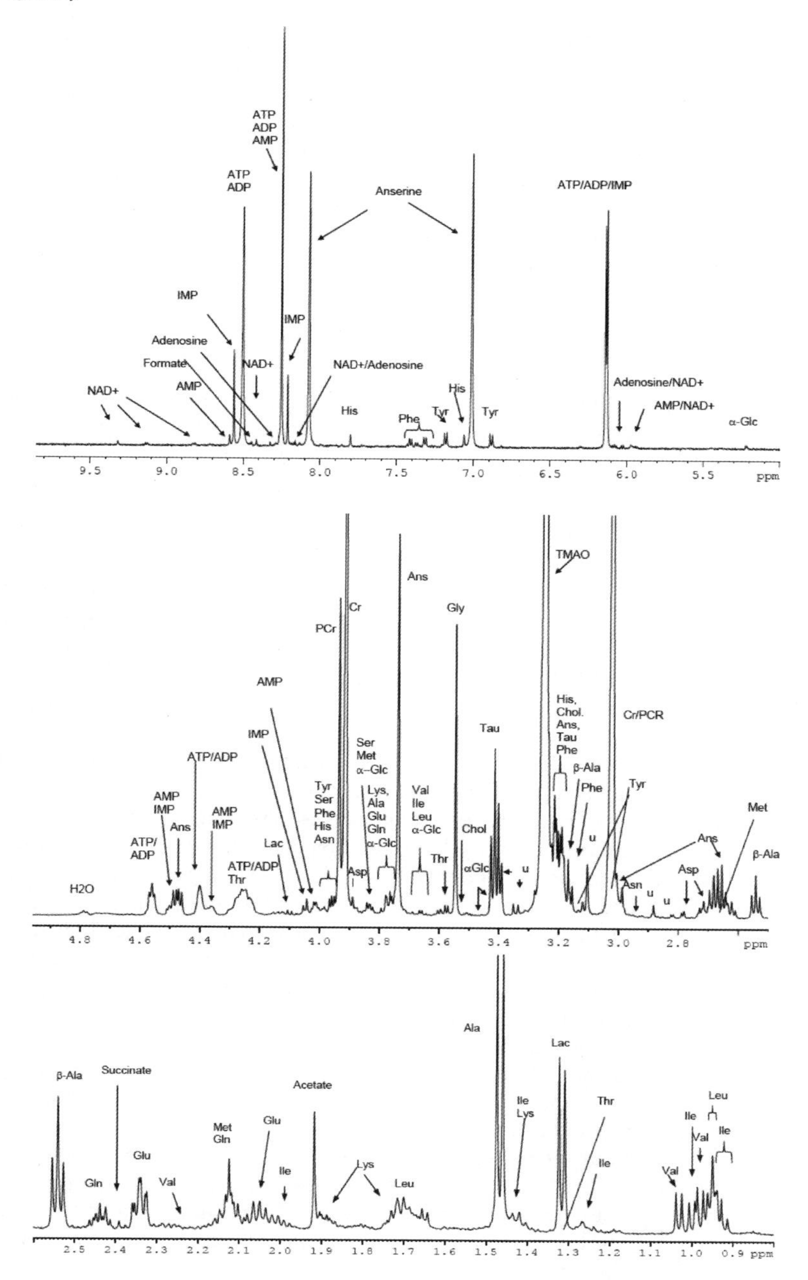

Figure 1 *Representative 500MHz ^{1}H NMR spectrum of PCA extracts (pH=7.2) from white muscle of cod.*

A wide range of metabolites was assigned, many of which give information on the metabolic condition of the sample analyzed. Differences both within and between the two fish species were observed. Relatively large differences between individuals of the same species were seen. This sample to sample variance may be due to the individual variation known to be found amongst fish samples[17] and differences in the handling of the fish (i.e. stress level)[18]. In addition, small inconsistencies in the extraction procedure might influence the level of labile compounds in the post-mortem fish muscle. Spectra of cod and haddock displayed to a large degree the same signals. This is consistent with earlier findings, which showed considerable overlap in chemical composition of the nitrogenous extractives of the two species[19,20].

3.2 Selected assigned metabolites

The peak of the absolute highest intensity in the spectra arises from the natural osmolyte trimethylamine N-oxide (**TMAO**). Many suggestions about the function of TMAO have been made. It is now generally believed that TMAO has an osmoregulatory role[21]. TMAO is reduced by bacteria to trimethylamine (TMA), which is responsible for the unpleasant "fishy" odor and taste of spoiled fish[22]. The signal arising from TMA (singlet at 2.91 ppm)[13] was not detected in the fish analyzed in our study, since the samples were taken immediately after death.

Creatine (Cr) and phosphocreatine (PCr) display singlets of high intensity in the spectra. Most of the Cr is phosphorylated (PCr) in resting muscle and supplies energy in the form of high energy phosphate for muscular contraction. Fish exposed to stress prior to death have lower values of PCr and ATP than unstressed fish[18]. The observed signal from PCr (3.92 ppm) in this study suggests that PCr has been preserved during the catching and the extraction procedure of the fish.

The signals from **amino acids** in the NMR spectra most likely stem from free amino acids since these are abundant in fish muscle; however, there might be contributions from peptides too. These compounds play important roles in physiological functions including osmoregulation and buffering capacity[21], in addition they contribute to the aroma and flavor of the fish[22] and increase its antioxidant capacity[23,24]. During storage and processing, proteins are degraded, and the level of free amino acids and peptides in muscle changes.

High content of **taurine** has been shown to be characteristic of white-fleshed fishes[22]. **Choline,** a precursor of acetylcholine, is important as a methyl donor in various metabolic processes including lipid metabolism[21]. Betaine, a derivative of choline, has been associated with osmoregulation[25] and acts as methyl donor in the synthesis of methionine from homocysteine. Previous studies showed that ^{1}H NMR could be used to quantify betaine in mussels[26] and cod[11]. In the present work betaine could not be detected because the two singlets arising from betaine (at 3.91 and 3.27 ppm), are presumably hidden under the higher intensity signals of creatine and TMAO.

Distinct peaks from **anserine** were visible both for cod and haddock samples. This compound is believed to function as a buffer during anaerobic metabolism, a fact reflected by the high levels in muscles used for burst activity[27]. In addition to an intracellular buffering function[28,29], anserine have been proposed to have additional roles controlling enzyme activity[30], inhibiting oxidative reactions[31] and as neurotransmitter[32]. Anserine decomposes to its constituents β-alanine and 1-methylhistidine by hydrolysis, a fact that permitted Ruiz-Capillas *et al.*[33] to estimate the loss of quality during ice storage by measuring the levels of 1-methylhistidine, β-alanine, anserine and tryptophan.

Adenosine triphosphate *(ATP)* predominates the **nucleotides** in muscle of live animals under normal conditions, but after death a series of enzymatic reactions leads to decomposition of ATP to adenosine diphosphate (*ADP*), adenosine monophosphate (AMP), inosine monophosphate (*IMP*), inosine (Ino) and hypoxanthine (Hx). When the level of ATP has dropped under a critical level; the muscle enters *rigor mortis*, however both the extent of rigor and the nucleotide level varies along the fish body[34]. Another important factor in analyzing ATP in muscle is the sample handling and extraction procedure[35,11]. Hx is a contributor to the bitter off-flavour of spoiled fish while IMP is usually associated to the desirable taste of fresh fish. The degree of freshness is often expressed as the K-value[36], an indication of the content of ATP relative to its degradation products. Sitter *et al.* showed that ^{1}H NMR of perchloric extracts is a possible method for studying ATP degradation, and estimated the loss of freshness and K-value during ice storage of Atlantic halibut (*Hippoglossus hippoglossus L*)[9]. However, in some species the degradation of nucleotides is too fast, as in Pacific cod (*Gadus morhua macrocephalus*), or too slow, as in plaice (*Paralichthys olivaceus*), for the K-value to be of practical use in quality control[37].

In the present work, AMP and IMP were unambiguously identified in the spectra, while ATP and ADP could not be distinguished due to overlapping peaks, both in the 1D and the 2D spectrum. IMP content of unstressed fish at slaughter is found to be low[38,34]. The relatively high level of IMP in the spectra indicates that the fish were stressed at the time of slaughter, while the absence of detectable signals from Hx in the spectra is in accordance with the freshness of the fish. In comparison Gribbestad *et al.* assigned hypoxanthine in the spectra of salmon muscle which had been stored on ice for more than two days[10].

The doublet arising from **lactate** in the 1D spectrum is well resolved. Previous studies have shown that the level of lactate in fish muscle can be calculated both from ^{1}H NMR and ^{13}C NMR spectra[39,10]. The lactate concentration reflects the initial glycogen stores before death, the handling of the fish and the extraction procedure.

3.3 Species differences: cod vs haddock

Examples of differences between cod and haddock are given in Figure 2, where peaks arising from β-alanine, glutamate and glutamine are all of lower intensity in haddock than in cod. Conversely, the signals from α-glucose and β-glucose were higher in haddock than in cod (results not shown). More samples need to be analyzed to make conclusions about whether these differences are due to species-specific differences or differences in postmortem handling of the fish, fishing grounds, season, sex, species, size or biological conditions.

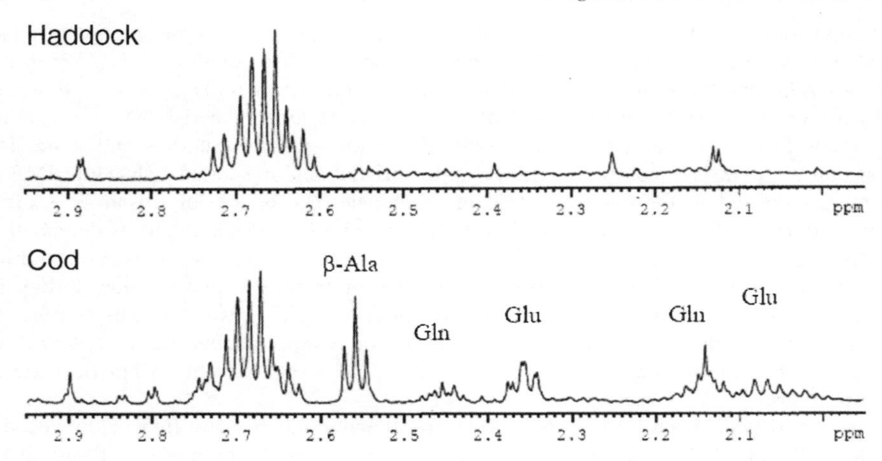

Figure 2 *Representative 500MHz ^{1}H NMR spectra of perchloric acid extracts of white muscle from haddock and cod. The lower level of β-Ala, Glu, and Gln in haddock compared to cod are examples of differences between the metabolic profiles obtained in this study.*

4 CONCLUSIONS

A wide range of metabolites was assigned, many of which give information about the metabolic condition of the sample analyzed. Cod and haddock displayed, to a large degree, the same compounds, however both intra- and interspecific differences in relative peak intensities were registered. Fish is known to display large individual differences in metabolite composition, and the present study does not permit one to attribute the differences to species specific factors or to others, such as environmental factors, development stage or physical condition. However, the results of this work may have practical applications in the use of high resolution NMR to further elucidate biochemistry and nutritional aspects of fish.

Abbreviations

Ile, isoleucine; Leu, leucine; Val, valine; Lac, lactate; Ala, alanine; Lys, lysine; Glu, glutamate; Met, methionine; Gln, glutamine; β-Ala, β-alanine; Ans, anserine; α-Glc, α-glucose; β-Glc, β-glucose; Asp, aspartate; Asn, asparagine; TMA, trimethylamine; Cr, creatine; PCr, phosphocreatine; Tyr, tyrosine; Phe; phenylalanine; Chol, choline; His, histidine; TMAO, trimethylamine N-oxide; Tau, taurine; Gly, glycine; Thr, threonine; Ser, serine; AMP, adenosine monophosphate; IMP, inosine monophosphate; ATP, adenosine triphosphate; ADP, adenosine diphosphate; NAD+, Nicotinamide adenine dinucleotide; u, unassigned peak.

Acknowledgements

The Norwegian Research Council is gratefully acknowledged by the financial support to NFR-project no. 146932/130 and 154137/130.

References

1 T. Arakawa and S. N. Timasheff, *Biophys. J.*, 1985, **47,** 411.
2 M. B. Burg, E. D. Kwon and D. Kültz, *Annu. Rev. Physiol.*, 1997, **59,** 437.
3 H. P. Redmond, P. P. Stapleton, P. Neary and D. Bouchier-Hayes, *Nutrition*, 1998, **14,** 599.
4 S. Schaffer, K. Takahashi and J. Azuma, *Amino Acids*, 2000, **19,** 527.
5 P. P. Stapleton, R. P. Charles, H. P. Redmond and D. J. BouchierHayes, *Clin. Nutr.*, 1997, **16,** 103.
6 W. C. Hou, H. Chen and Y. H. Lin, *J. Agric. Food Chem*, 2003, **51,** 1706.
7 S. T. Chambers, B. A. Peddie, K. Randall and M. Lever, *Int. J. Antimicrob. Agents*, 1999, **11,** 293.
8 N. Howell, Y. Shavila, M. Grootveld and S. Williams, *J. Sci. Food Agric.*, 1996, **72,** 49.
9 B. Sitter, J. Krane, I. S. Gribbestad, L. Jørgensen and M. Aursand, in *Advances in Magnetic resonance in Food Science,* eds. P. S. Belton, B. P Hills, G. A. Webb, Royal Society of Chemistry, Cambridge, 1999, p. 226.
10 I. S. Gribbestad, M. Aursand and I. Martinez, *Aquaculture*, 2005, **250,** 445.
11 I. Martinez, T. Bathen, I. B. Standal, J. Halvorsen, M. Aursand, I. S. Gribbestad and D. Axelson, *J. Agric. Food Chem*, 2005, **53,** 6889.
12 T. Glonek, S. J. Kopp, E. Kot, J. W. Pettegrew, W. H. Harrison and M. M. Cohen, *J. Neurochem*, 1982, **39,** 1210.
13 S. Maschke, A. Wahl, N. Azaroual, O. Boulet, V. Crunell, M. Imbenotte, M. Foulard, G. Vermeersch and M. Lhermitte, *Clin. Chim. Acta*, 1997, **263,** 139.
14 J. K. Nicholson and P. J. D. Foxall, *Anal. Chem.*, 1995, **67,** 793.
15 B. Sitter, T. Bathen, B. Hagen, C. Arentz, F. E. Skjeldestad and I. S. Gribbestad, *Magn. Reson. Mater. Phy.*, 2004, **16,** 174.
16 T. W. M. Fan, *Prog. Nucl. Magn. Reson.*, 1996, **28,** 161.
17 I. Martinez, B. Dreyer, A. Agersborg, A. Leroux, and G. Boeuf, *Comp. Biochem. Phys. B*, 1995, **112,** 717.
18 Erikson U, A. R. Beyer and T. Sigholt, *J. Food Sci.*, 1997, **62,** 43.
19 J. M. Shewan, L. I. Fletcher, S. M. Partridge and R. C. Brimley, *J. Sci. Food Agric.*, 1952, **3,** 394.
20. J. M. Shewan, *J. Sci. Food Agric.*, 1953, **4,** 565.
21. A. Van Waarde, *Comp. Biochem. Physiol. B*, 1988, **91,** 207.
22. S. Konosu and K. Yamaguchi, in *Chemistry and Biochemistry of Marine Food Products,* eds. R. E. Martin, G. J. Flick, C. E. Hobart, D. R. Ward, AVI publishing, Westport, 1982, p. 367.
23 K. M. Chan and E. A. Decker, *Crit. Rev. Food Sci. Nutr.*, 1994, **34,** 403.
24 H. Wu, H. Chen and C. Shiau, *Food Res. Int.*, 2003, **36,** 949.
25 S. K. Pierce, L. M. Rowlandfaux and B. N. Crombie, *J. Exp. Zool.*, 1995, **271,** 161.
26 C. G. N. de Vooys and J. A. J. Geenevasen, *Comp. Biochem. Physiol. B,* 2002, **132,** 409.
27 A. A. Boldyrev and S. E. Severin, *Adv. Enzyme Regul.*, 1990, **30,** 175.
28 H. Abe, G. P. Dobson, U. Hoeger, and W. S. Parkhouse, *Am. J. Physiol.*, 1985, **249,** R449.
29 H. Abe, *Biochemistry (Mosc.)*, 2000, **65,** 891.
30 T. Ikeda, K. Kimura, T. Hama and N. Tamaki, *J. Biochem.*, 1980, **87,** 179.
31 A. A. Boldyrev, A. M. Dupin, E. V. Pindel and S. E. Severin, *Comp. Biochem. Physiol. B*, 1988, **89,** 245.

32 A. Boldyrev, *Trends Pharmacol. Sci.*, 2001, **22,** 112.
33 C. Ruiz-Capillas and A. Moral, *Eur. Food Res. Technol.*, 2001, **212,** 302.
34 Cappeln G and F. Jessen, *J. Food Sci.*, 2002, **67,** 991.
35 G. Cappeln, J. Nielsen and F. Jessen, *J. Sci. Food Agric.*, 1999, **79,** 1099.
36 I. Karube, H. Matsuoka, S. Suzuki, E. Watanabe and K. Toyama, *J. Agric. Food Chem.*, 1984, **32,** 314.
37 S. Ehira, *Bull. Tokai Reg. Fish. Res. Lab.*, 1976, **88,** 1.
38 T. Berg, U. Erikson and T. S. Nordtvedt, *J. Food Sci.*, 1997, **62,** 439.
39 M. Aursand, L. Jørgensen and H. Grasdalen, *Comp. Biochem. Physiol. B*, 1995, **112,** 315.

NMR OF CELL WALLS : A MULTI-SCALE APPROACH

C. Rondeau-Mouro, H. Bizot and M. Lahaye

UR Biopolymères, Interactions, Assemblages-INRA, Rue de la Géraudière, BP 71627, 44316 Nantes cedex 3, France

1 INTRODUCTION

Among their multiple functions, plant cell walls provide a protective barrier for cells against biotic and abiotic stresses and contribute to the mechanical resistance and morphogenesis of organs. They are composed of different polysaccharides (cellulose, hemicelluloses and pectins), proteins and polyphenols, which content, structure and organisation depend on the cells age, function as well as plant origin.[1] Representing only about 2% of the total mass of growing turgid tissues and in fleshy fruits, their major contribution to cell adhesion and organ mechanical properties impacts the different uses of crops. In particular, they play important roles on fleshy fruit texture, which variability represents an important part of the organoleptic characteristics orienting consumers' choice. It also conditions their storage and food processing ability. Our approach aims at characterizing molecular structures and organisations in cell walls, which determine texture of fleshy fruits. The tomato pericarp is selected as a typical model but additional insights are taken from the classical plant model *Arabidopsis thaliana* in order to identify pertinent genes involved in the cell wall biosynthesis and having an effect on wall rigidity and/or cell adhesion. Search for natural mutants of tomato is also considered for recognition of genetic markers of texture. Another approach is used based on the reconstruction of biomimetic systems from bacterial cellulose produced by *Acetobacter xylinus* or from in vitro re-assemblies of isolated wall polysaccharides. We illustrate different NMR techniques to investigate the cell wall assemblies using a multiscale strategy. The molecular level is described from the compositional and structural points of view, while the nano-microtexture scale is respectively tackled by proton relaxometry (T_2 and $T_{1\rho}$) of confined water or polysaccharides.

2 MATERIALS AND METHODS

2.1 NMR spectroscopy

Solid-state NMR experiments were performed on a Bruker DMX-400 spectrometer operating at a ^{13}C frequency of 100.62 MHz and equipped with a double resonance H/X CP-MAS 4mm probe. The MAS rate was fixed at 7000 Hz and each experiment was recorded at ambiant temperature (294 ± 1 K). The CP-MAS pulse sequence used a 4 μs 90° proton pulse, a 1ms contact time at 62.5 kHz and a 7 s recycle time preceding an acquisition time of 52 ms during which dipolar decoupling (TPPM[2]) was applied. Chemical shifts were referenced to glycin, assigning the carbonyl carbon at 176.03 ppm. Measures of the proton rotating-frame relaxation times $T_{1\rho}$ were achieved using a delayed-contact experiment by varying the spin-locking pulse delay according to Newman and Hemmingson.[3] Relaxometry measurements were performed on the final rehydrated cell wall residues using a low-field NMR spectrometer operating at 20 MHz for proton (Minispec Bruker PC 120, 0.47T). The spin-spin relaxation times (T_2) were measured from the free induction decay (FID) and the Carr Purcell Meiboom Gill (CPMG) sequence with a recovery delay of 5 s, a delay between the 90° and 180° pulses of 0.1 ms using 3000 points for acquisition. The relaxation decay curves were adjusted by the Maximum Entropy method (MEM[4]).

2.2 Preparation of cellulose-xyloglucans composites and cell walls residues

The *Arabidopsis Thaliana* plant samples were obtained from partner labs of INRA (Vegetal Biology in Versailles). Xyloglucan Glyloid was a gift from Dainippon-Pharm Co. (YAMATOYA K. 2003). Bacterial cellulose was produced following the well established methodology of Atalla & al.[5] using the same patented strain of Gluconacetobacter Xylinus (ATCC 53524) propagated in H&S Medium[6]. In order to minimize the desorption of incorporated hemicellulose, a proteinase purification scheme was preferred[7] here based on Sigma-Aldrich Proteinase K (0.01 mg / mg Protein) in SDS 1% at pH8 (Tris HCl) for 3h at 37°C. The level of high molecular weight (650 kD) xyloglucan in the growth media was limited to 0.5%. Static culture in Roux Flasks developed for 5 days at 25°C from an 0.5%v.v. pelletized innoculum. Upon XG incorporation the texture of the roughly 1% solid fleece changed from a tenacious mat to a very poorly cohesive network as described by Astley & al. 2002). Never dried but comminuted samples were blotted on non cellulosic napkins (Crew 2, Kimberly-Clark) before squezzing in SS-NMR rotors (%H_2O d.b. = 100-120%.). Cell walls were prepared as alcohol insoluble residues (AIR) as described by Bouton et al.[8] Removal of pectins, starch and proteins followed sequential extractions and enzymatic treatments. Typically 500mg of AIR in a phosphate buffer (0.2M, pH 7.2) were submitted to the following series of extractants : i) NaCl 0.5M and Triton x-100 0.5% , ii) potassium oxalate (1 mg/ml), iii) NaCl 0.5M, and (iv) DMSO. Successive residues were collected by centrifugation at 28600 g for 15 min at 20°C and starch was eliminated according to Prosky et al.[9] Proteins were removed by acidified 80% phenol w/w.[10] Further pectins and proteins removal relied on enzymatic treatments, with a laboratory purified endo-polygalacturonase II (Novo T-94497) followed by protease (Sigma, *Bacillus licheniformis*, p-5380), trypsin (Sigma-Aldrich, t-8802), proteinase K (Merck, *Tritirachium album limber*, p-6566) and Pronase (Fluka, *Streptomyces griseu;s*). Finally, rhamnogalacturonase II (NOVO, PPJ4478) and endo-polygalacturonase II as above were used to remove residual pectins. Residues at each step were recovered by centrifugation (14000g, 15 min) and rinsed with 0.5M NaCl and deionized water.

3 RESULTS and DISCUSSION

3.1 Nano- and meso-scale investigations of polysaccharides interactions

Solution NMR spectroscopy is very efficient for the structural characterization of soluble polysaccharides[11-17]. However, in intact cell walls, their characterization is achieved by solid-state NMR with cross-polarisation (CP) coupled with magic angle spinning and high-power proton decoupling in order to reduce chemical shift and dipolar anisotropic interactions inherent to solid materials[18-22]. Application of the CPMAS technique (sequence in Figure 1) for investigating structures and conformations of cell wall components has been demonstrated since 1984 with the discovery of the two cellulose allomorphs Iα and Iβ by Atalla and Vanderhart.[23] The Iα form corresponds to a single-chain triclinic crystallographic symmetry, whereas Iβ is monoclinic and characterised by two parallel chains. Native celluloses generally appear as a mixture of these two forms, the ratio between Iα and Iβ depends on the source and to a lesser extent to environmental conditions. It was also shown that the Iα phase is metastable and could be converted into the Iβ form by a hydrothermal treatment at 260°C.[24] This first high-resolution NMR study of cellulose study remains a reference work for cellulose characterisation, since crystalline level and microfibrils lateral size can be estimated by NMR spectral decomposition. At the same time, measures of relaxation times on dried apple cell walls were realized by Irwin et al.[25]

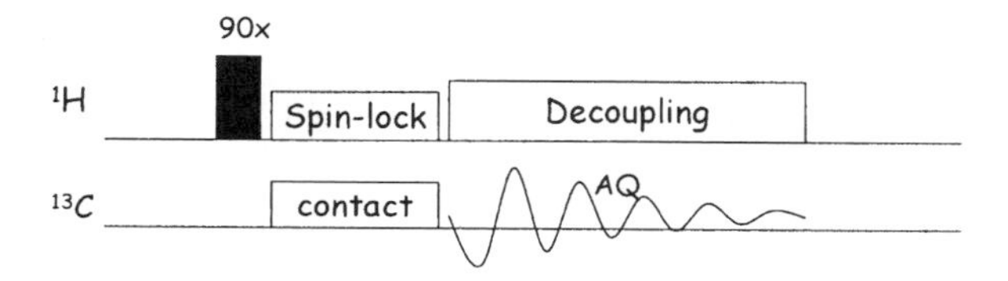

Figure 1 *Cross-polarisation magic angle (CPMAS) sequence for carbon detection*

Hydrated cell walls mobility resolved techniques which exploit relaxation time contrasts between locally vicinal phases have permitted to investigate domains of few nanometers using protons and carbons relaxation times experiments modulated by proton spin diffusion (for spatially adjacent nuclei).[26-29] At this scale, interactions between the cell wall constituents can be characterised. One of these assemblies, still not totally understood, concerns the intimate interactions between cellulose and xyloglucans, the major hemicellulose in the primary cell walls of non-graminaceous plants. These interactions are inferred from co-extraction of cell wall components[30,31], from microscopic observations[32] and from *in-vitro* sorption experiments of isolated xyloglucans on cellulose[33]. Our approach consisted in reconstructing a cellulose-xyloglucans composite in order to address the intimate association as related to adsorption of xyloglucans on nascent cellulose e.g. occurring between the primary crystallization and the subsequent fasciation of larger ribbons. Addition of tamarind xyloglucans to the culture medium of the *Acetobacter xylinus* bacterium permitted to produce the composite. Figure 1A displays the CPMAS spectrum of bacterial cellulose in a hydrated state showing signals at 105 ppm, 88-91 ppm, 84-86 ppm assigned respectively to C1 sites, crystalline and amorphous C4 carbons. The C2, C3 and C5 carbons resonate between 69 and 78 ppm while the C6 carbon shows resonances between 57-64 ppm (amorphous chains) and 64-67 ppm (crystalline chains). These peaks are clearly present in the figure 1C showing the carbon spectrum of the hydrated pellicle produced in the presence of xylogucans. The larger bandwidths of the cellulose signals in addition to the decrease of the C1 components centred at 105 ppm, have been correlated to the expected conversion of the Iα allomorph to the Iβ allomorph.

The occurrence of xyloglucans is detected at 99.1 ppm, by the specific signal of xylose. For comparison, CPMAS spectrum of hydrated tamarind xyloglucans is presented in figure 1B.

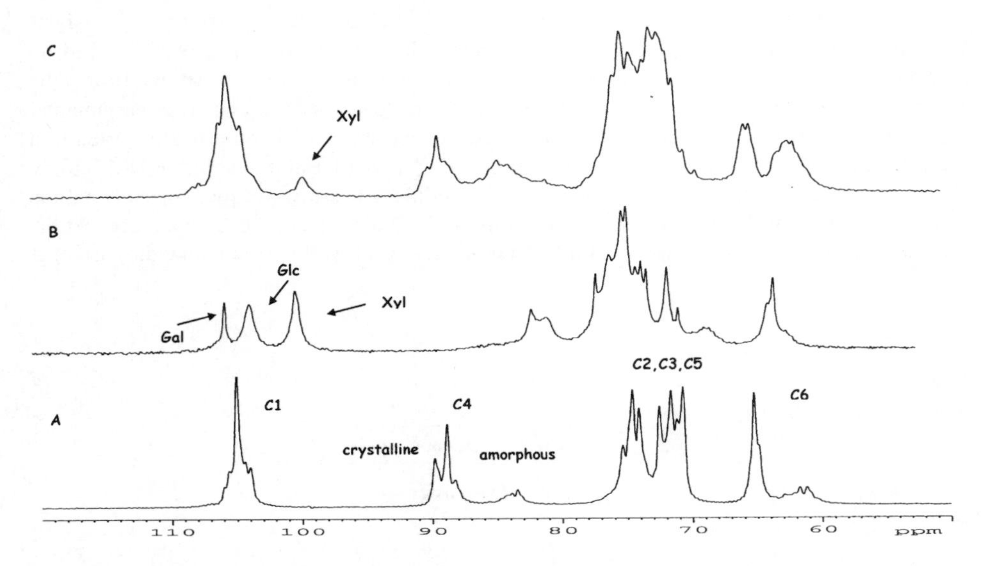

Figure 2 *CPMAS spectra of A- cellulose synthesised with the Acetobacter xylinum bacterium, B- tamarind xyloglucans, C- composite of bacterial cellulose-xyloglucans. Gal: galactose, Glc: glucose, Xyl: xylose*

Motions of specific nuclei can be probed by measuring their relaxation times spin lattice in the rotating frame $T_{1\rho}$ and spin/spin T_2[26-29]. While spin-lattice relaxation in the laboratory frame relaxation times T_1 inform about fast molecular motions with characteristic frequency in the MHz range (10^8-10^{11} Hz) corresponding to side groups movements as methyl esters for example, spin-lattice relaxation in the rotating frame $T_{1\rho}$ gives access to slower motions as movements of the polymer chains in the kHz range (10^4-10^6 Hz). Measurements of protons T_2 and $T_{1\rho}$ relaxation times are very relevant for the analysis of the nano-mechanical properties of cell walls and provide means to study interactions between polysaccharides as well as mechanical changes of cell walls at a nano-scale level. These can be achieved using the cross-polarisation technique through the carbons acquisition introducing a proton spin-lock or a delay for $T_{1\rho}$ and T_2 respectively, before the contact time (see figure 1).[3,34] An example of such experiments is illustrated on Figure 3 which displays the protons $T_{1\rho}$ values on isolated polysaccharides for a given hydration level, as well as for the pellicle produced by the *Acetobacter xylinum* bacterium in the presence of tamarind xyloglucans. Results clearly show that protons of flexibly (1,6)-linked xyloses in isolated xyloglucans are very short, around 1 ms, while when interacting closely within the pellicle, xylose protons display two relaxational components, with the longer one (27 ms) being very close to the proton $T_{1\rho}$ value measured on glucoses of cellulose in the pellicle (32 ms). This result reveals that xyloglucans are associated to bacterial cellulose fibrils. Further works are in progress in order to precisely characterize the adsorption mode and the interaction type.

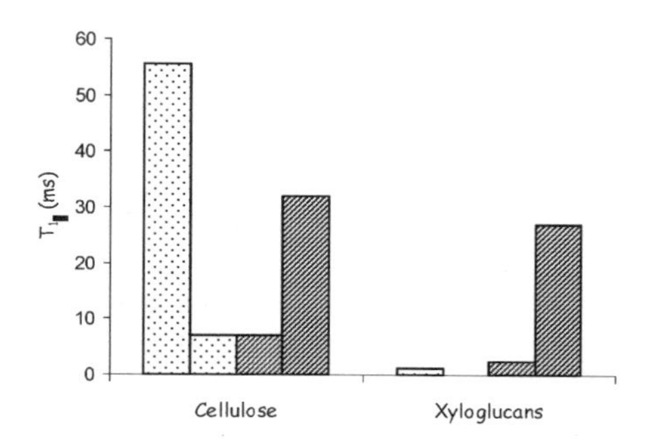

Figure 3 *Illustration of the proton $T_{1\rho}$ of isolated polysaccharides cellulose and xylogucans ... , and $T_{1\rho}$ of each polysaccharide inside the composite ///*

3.2 Nano- to macro-scale investigations of porosity of cell walls

Cell walls can be viewed as porous media with three-dimensional distributed solid interface consisting of stretched spheres representing isolated chains or composites of polysaccharides and/or structural proteins (Figure 4). Porosity is in fact the liquid volume over total sample volume. Rigidity of cell walls is mainly associated with the cellulose content and organisation but its extensibility was shown to be dramatically altered by deposition of pectins or xyloglucans.[33,36] Microscopic investigations of these re-constructed composites clearly indicated a regulation of pectins or xyloglucans on the cellulose three-dimensional organisation.

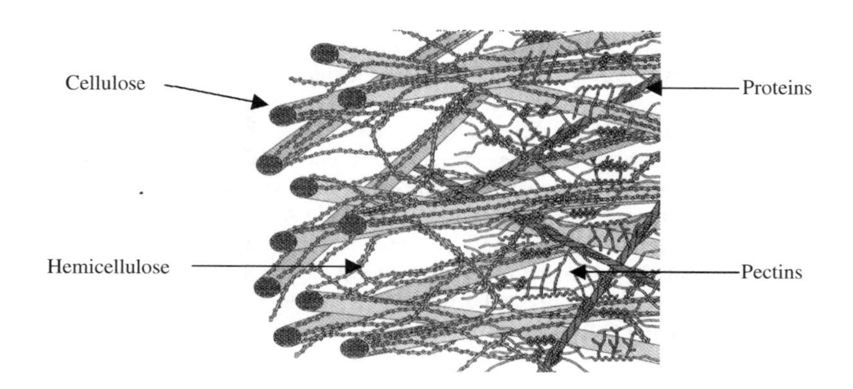

Figure 4 *Architecture model of primary cell walls from Carpita and MacCann*[35]

As a consequence, porosity of cell walls should modulate the cell strength and rigidity, so contributes to the textural perception of fleshy fruits. Hydration properties of cell walls are also dependant of their porosity as well as enzymes accessibility and reactivity in walls.

Cell walls are known to be rather heterogeneous in composition, depending on the plant origin and evolving in composition and architecture with time[1,35]. Among the various cell walls constituents, pectins have been shown to strongly contribute to the cell walls porosity.[37,38] Indeed, pectins form hydrated gels or three-dimensional networks through ionic junctions via Ca^{2+} between neighboring charged pectin carboxyl groups (COO-).[1,35,39]

In order to estimate the influence of pectins in the porosity regulation of cell walls, we used the *Arabidopsis thaliana* mutant *QUA1* (quasimodo) known to be affected on a gene that codes for a family 8 glycosyl transferase attributed to a galacturonosyl transferase involved in homogalacturonans synthesis.[8,40] Homogalacturonans (HG) are pectic segments composed of galacturonic acids more or less methyl- and acetyl-esterified. *QUA1* plants and cell suspension cultures showed a deficience in HG which were implicated in a lower proportion of calcium cross-linkages in the cell junction areas.[8,40,41] Being non-invasive, different complementary NMR methods can measure three-dimensional transport events between the molecular scale (nanometric scale by relaxometry) and the tissue scale (millimetric scale by imaging techniques). Water relaxometry measured on water protons inside cellulose fibres allowed estimation of an average pore size[42,43]. Moreover, it has been shown by Hills[44] that for aqueous biopolymer systems, the contribution of hydration water ("bulky") to the global water proton relaxation is minor compared with the dominant effects of fast exchanges between water and biopolymers protons. As a consequence, measurements of the water relaxation times in biopolymer systems allow describing surface diffusion phenomenon of water at the interface of the bulky water and the solid phase composed of cell wall polysaccharides. Distribution of these relaxation times is directly related to the pore distribution of the samples.[44,45]

For the first time, the water protons T_2 relaxation times have been measured on cellulose-hemicelluloses residues extracted from stems and microcalli of wild type *Arabidopsis thaliana* (WS and WC respectively) and others obtained from stems of the *Arabidopsis thaliana* mutant *QUA1* (QS). Figure 5 presents the T_2 distributions obtained for the three residues WS, WC and QC. The *QUA1* mutant of stems residues (QS) showed three major distributions centred at 2.8, 27.8 and 70.8 ms, while the wild type stems residues (WS) display a larger continuous distribution with maxima at 12.5, 37.2 and 102.3 ms.

The T_2 distribution of microcalli residues (WC) is weaker and mainly centred at 55.5 ms with a shoulder at 27.8 ms and a thin peak at 2.2 ms. This distribution is very similar to the *QUA1* stems ones.

These results indicate a rather wide heterogeneity of pore size for wild type stems cell walls residues (WS). Comparison of values measured on wild-type residues, indicated that stems residues of cell walls displayed wider pore sizes than for microcalli. This result should be correlated with the presence in stems cell walls of a secondary layer known to be composed of differently oriented cellulose and containing aromatic components[35]. This agrees with the idea that porosity should change during the wall development as plant growths.[46-49] Cell walls residues extracted from stems of the *QUA1* mutant showed a weaker heterogeneity and porosity than for wild type stems. While the homogalacturonan domains of pectins are known to form three-dimensional networks through calcium junctions[1,35,39], it is clear that their deficiency influences the cellulose-hemicellulose organisation. Quantification of neutral sugars coupled with liquid-state NMR structural analysis of pectins and hemicelluloses should help at understanding the *QUA1* mutation effects on cell walls porosity. Additional self diffusion measurements of molecular probes have also been realised for the description of the cell walls microgeometry (submitted elsewhere).

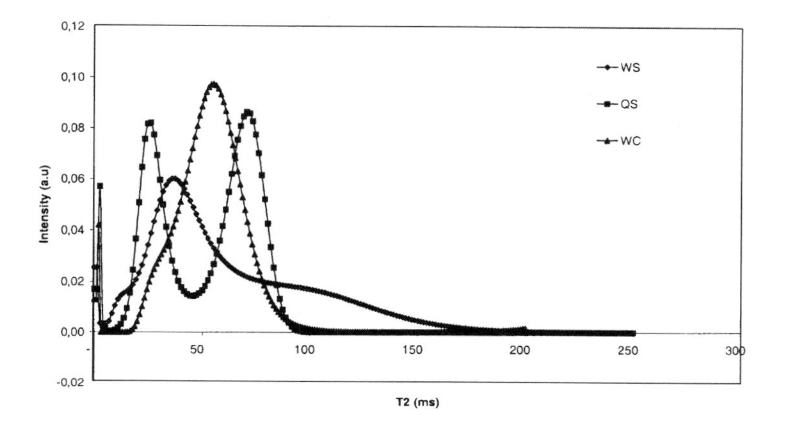

Figure 5 *Water proton T_2 distributions obtained by a MEM treatment (see Materials and Methods) for residues prepared from stems of wild type Arabidopsis thaliana (WS), from microcalli of wild type Arabidopsis thaliana (WC) and from stems of the Arabidopsis thaliana mutant QUA1 (QC).*

4 CONCLUSIONS

Plant cell walls are sophisticated and highly hydrated nanostructures able to withstand mechanical strains due to its cellulosic composition and composite nature. Interactions between cellulose, hemicelluloses and pectins seem to widely influence the mechanical properties of cell walls. Among the various techniques used to characterise cell walls, NMR spectroscopy helps at analysing the constituents' structure and associations at nanometric scales. While solid-state NMR is used to identify strongly rigid materials mainly composed of the cellulosic phase, liquid-state NMR reveals flexible polymer chains and water motions inside cell walls. A new approach consisting in the pore size estimation is considered aimed at understanding capillary forces, enzymes activity as well as textural properties of cell walls. In complement to the water protons relaxometry methods giving access to surface diffusion in nanometric pores, the self-diffusion coefficient measurements of molecular probes should permit investigation of volume transport events in few microns across.

5 AKNOWLEDGEMENTS

The authors thank J. Vigouroux, J. Davy and M. Cambert for their technical assistance respectively in the extraction of plant cell walls, in the cellulose cultures and in the low-field NMR measurements.

References

1 N.C. Carpita and D.M. Gibeaut, *Plant Journal*, 1993, **3**,1.
2 A.E. Bennett, C.M. Rienstra, M. Auger, K.V. Lakshmi and R.G. Griffin, *J. Chem. Phys.*, 1995, **103**, 6951.
3 R.H. Newman and J.A. Hemmingson, *Holzforschung*, 1990, **44**, 351.
4 F. Mariette, J.P. Guillement, C. Tellier and P. Marchal, In: D.N. Rutledge (eds), Signal treatment and signal analysis in NMR, Elsevier, Paris,1996, pp. 218.
5 R. H. Atalla, J.M. Hackney, I. Uhlin and N.S. Thompson, *Int. J. Biol. Macromol.*, 1993, **15** (2), 109.
6 S. Hestrin and M. Schramm, *Biochem. J.*, 1954, **58**, 345.
7 E. Roberts, I. Saxena and R. Brown, In: C. Schuerch (ed.), Cellulose and Wood – Chemistry and Wood – Chemistry and Technology. John Wiley and Sons, Inc., New York, 1989, pp. 689.
8 S. Bouton, E. Leboeuf, G. Mouille, M.T. Leydecker, J. Talbotec, F. Granier, M. Lahaye, H. Höfte and H.N. Truong, *Plant Cell*, 2002, **14**, 2577.
9 L. Prosky, N.G. Asp, T.F. Schweizer, J.W. De Vries and I. Furda, *J. Assoc. Off. Anal. Chem.*, 1988, **71**, 1017.
10 S.C Fry, The Growing Plant Cell Wall: Chemical and Metabolic Analysis. Longman Scientific and Technical, New York, 1988, pp 333.
11 A.M. Gil, I.F. Duarte, I. Delgadillo, I.J. Colquhoun, F. Casuscelli, E. Humpfer and M. Spraul, *J. Agric. Food. Chem.*, 2000, **48**, 1524.
12 Q.W. Ni and T.M. Eads, *J. Agric. Food. Chem.*, 1993, **41**, 1026.
13 Q.W. Ni and T.M. Eads, *J. Agric. Food. Chem.*, 1993, **41**, 1035.
14 W.S. York, H. Van Halbeek, A.G. Darvill and P. Albersheim, *Carbohyd. Res.*, 1990, **200**, 9.
15 S. Levy, W.S. York, R. Stuike-Prill, B. Meyer and L.A. Staehelin, *Plant J.*, 1991, **1**, 192.
16 P. Ryden, I.J. Colquhoun and R.R. Selvendran, *Carbohyd. Res.*, 1989, **185**, 233.
17 S. Cros, C. Herve du Penhoat, N. Bouchenal, H. Ohassan, A. Imberty and S. Perez, *Int. J. Biol. Macromol.*, 1992, **14**, 313.
18 C.M.G.C. Renard and M.C. Jarvis, *Plant Physiol.*, 1999, **119**, 1315.
19 M.C. Jarvis, *Phytochemistry*, 1994, **35**, 485.
20 M.C. Jarvis and D.C. Apperley, *Plant Physiol.*, 1990, **92**, 61.
21 M.C. Jarvis, *Carbohyd. Res.*, 1990, **201**, 327.
22 K.M. Fenwick, D.C. Apperley, D.J. Cosgrove and M.C. Jarvis, *Phytochemistry,* 1999, **51**, 17.
23 R.H. Atalla and D.L. Vanderhart, *Science*, 1984, **223**, 283.
24 H. Yamamoto, F. Horii and H. Odani, *Macromolecules*, 1989, **22**, 4130.
25 P.L. Irwin, P.E. Pfeffer, W.V. Gerasimowicz, R. Pressey and C.E. Sams, *Phytochemistry*, 1984, **23**, 2239.
26 H.R. Tang, Y.L. Wang, P.S. Belton, *Solid State N.M.R.,* 2000, **15** (4), 239.
27 T.J. Foster, S. Ablett, M.C. McCann, M.J. Gidley, *Biopolymers*, 1996, **39** (1), 51.
28 M.A. Ha, D.C. Apperley and M.C. Jarvis, *Plant Physiol.*, 1997, **115**, 593.
29 M.C. Jarvis, K.M. Fenwick and D.C. Apperley, *Carbohyd. Res.*, 1996, **288**, 1.
30 W.S. York, A.G. Darvill, M. McNeill, T.T. Stevenson and P. Albersheim, *Methods Enzymol.*, 1985, **118**, 3.
31 L.D. Talbott and P.M. Ray, *FEBS Lett.,* 1992, **98**, 357.
32 M.C. McCann, B. Wells and K. Roberts, *Journal of Cell Science*, 1900, **96**, 323.

33 E. Chanliaud, J. De Silva, B. Strongitharm, G. Jeronimidis and M.J. Gidley, *Plant Journal*, 2004, **38**, 27.
34 P. Tekely and M.R. Vignon, *J. Polym. Sci. Part C Polym. Lett.*, 1987, **25**, 257.
35 N. Carpita and M.C. McCann. In: B. Buchanan, W. Gruissem, R. Jones (eds) Biochemistry and molecular biology of plants. American society of plant physiologists, 2002, pp 52.
36 E. Chanliaud and M.J. Gidley, *Plant Journal*, 1999, **20**, 25.
37 O. Baron-Epel, P.K. Gharyal and M. Schindler, *Planta*, 1988, **175**, 389.
38 A. Fleischer, M.A. O'Neill and R. Ehwald, *Plant Physiol.*, 1999, **121**, 829.
39 M.C. McCann and K. Roberts. In: C.W. Lloyd (eds) The cytoskeletal basis of plant growth and form. Academic press, London, 1991, pp 109.
40 C. Orfila , S.O. Sorensen, J. Harholt, N. Geshi, H. Crombie, H.N. Truong, J.S.G. Reid, J.P. Knox and H.V. Scheller, *Planta*, 2005, **222**, 613.
41 E. Leboeuf, B. Thoiron and M. Lahaye, *J. Exp. Bot.*, 2004, **55**, 2087.
42 T.Q. Li, Interaction between water and cellulose fibres. Ph. Thesis, Royal institute of technology, Stockholm, Sweden, 1991.
43 B. Andreasson, J. Forsström, L. Wagberg, *Cellulose*, 2003, **10**, 111.
44 B.P. Hills, *Molecular Physics*, 1992, **76**, 489.
45 B.P. Hills, *Molecular Physics*, 1992, **76**, 509.
46 S. Amino, Y. Takeuchi and A. Komamine, *Physiol. Plant*, 1984, **60**, 326.
47 T. Asamizu, N. Nakano and A. Nishi, *Planta*, 1983, **158**, 166.
48 K. Ruel, G. Chambat and J.P. Joseleau. In : M.M.A. Sassen, J.W.M. Derksen, A.M.C. Emons, A.M.C. Wolters-Arts (eds) Book of abstracts of the sixth cell wall meeting. University Press, Nijmegen, 1992, pp 145.
49 M.C. McCann and K. Roberts, *J. Exp. Bot.*, 1994, **45**, 1683.

MRI OF A MEAT-RELATED FOOD SYSTEM

J.P. Renou, J.M. Bonny, L. Foucat and A. Traoré

QuaPA/STIM, INRA Theix, St-Genès Champanelle, 63122, France

1 INTRODUCTION

NMR has undergone spectacular development in various branches of science and medicine. One of the reasons for this success is its ability to provide information non-destructively. MRI is now widely used in diagnostic radiology. Research is being conducted into MRI methods for clinical applications to increase diagnostic accuracy and decrease examination time. Despite having very different objectives, food science can also benefit from these developments in MRI.

Product quality depends on safety, technological characteristics, and sensory and nutritional values. Geographic origin and animal welfare fall more under the scope of fair trade. Technological characteristics such as texture, water-holding capacity (WHC) and pH are very important factors for both the meat processor and - via the product - the consumer. The meat supply chain consists of a series of stages in the course of which the animal carcass is progressively transformed into meat and meat products. Technological processes tailored to raw meat quality can lead to improvements in the quality of the product consumed. Although sensory qualities vary according to the consumer, there are physical measurements that can objectively gauge aroma, colour, toughness and juiciness. Recent health scares in the food industry - dioxin-contaminated poultry, BSE, illegal pesticide residues and avian influenza - have heightened consumer awareness in all aspects of health quality, in addition to microbiological risks such as Lysteria. Nutritional quality involves proteins, fats and micronutrients whose degradation and availability in the human body are dependent on technological processes. For example, oxidation, pH and temperature affect lipolysis, proteolysis, peptide aggregation, etc.

NMR is widely used as a means of characterizing food products. Food safety and nutritional aspects can be studied by metabolomics. This paper is limited to MRI and its application in the study of technological and sensory characteristics (Figure 1).

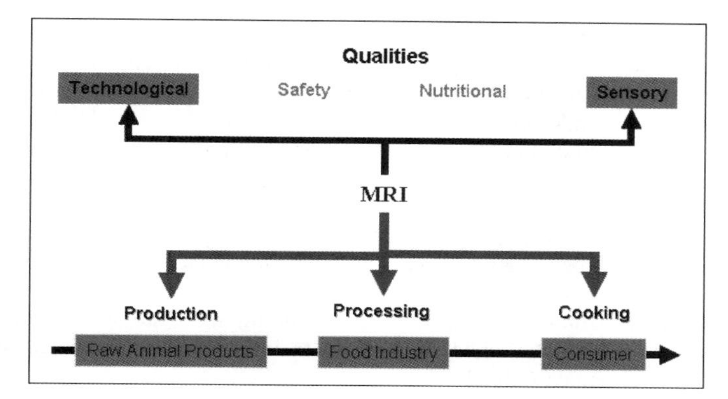

Figure 1 *MRI and meat qualities*

2 APPLICATIONS

2.1 Carcass classification

It is extremely useful to be able to assess technological quality as soon as possible after slaughter. There is a large body of studies dealing with MRI for the characterization of pig and broiler carcass composition[1-4]. In MRI procedures, the carcass can be placed into the magnet, and fat and lean meat are quantitatively determined using spatial information. A range of methodological adaptations are performed to increase the contrast between lean and adipose tissue and improve data processing. A linear relationship has been demonstrated (R^2 = 0.982) between dissected fat tissue weight and number of fat tissue voxels detected in MRI images of pig belly[5]. Based on this previous study, low-field MRI was used to predict pig carcass composition. In 4 primary cuts (ham, loin, shoulder and belly), MRI made it possible to predict muscle content, total fat and subcutaneous fat in the carcass. The best results were obtained from these four cuts with R^2 ranging from 0.918 to 0.997. The lowest accuracy was found for fat and especially subcutaneous fat [4] All these studies underlined that the MRI measurements give accurate quantifications for different tissues closely matching the values produced by dissection and chemical analysis. Given these results, MRI could be used as a reference method. However, its use as an on-line sensor in the abattoir is not yet a practical reality. The main problem resides in the sensitivity of NMR. The NMR scan takes several minutes or even hours, a duration that is incompatible with slaughtering rate. In the abattoir, one pig is killed every four seconds, one turkey every two seconds and two hens every one second. Other visioning techniques such as optical and ultrasound sensors can be used for carcass classification, but MRI can give a better estimation by improving the positioning of these sensors [2].
These industrial constraints mean that MRI is not a realistic option as an on-line sensor, but it remains useful as a reference tool as well as dissection method.

2.2 WHC

The interactions between water and macromolecules determine the Water Holding Capacity (WHC) of meat. Meat WHC depends primarily on the extent of *post mortem*

myofibrillar shrinkage and the correlative changes in the water compartments [6;7]. The WHC of fresh meat was assessed by NMR relaxation measurement of water protons [8;9]. large number of NMR studies have been performed over the last 20 years, and some studies have been conducted to assess meat quality [10,11,12,13.] Highly significant relationships were found between NMR relaxation parameters and certain other characteristics such as pH, cooking yield, etc [14]. In vivo, the muscle is organised in response to its physiological environment. Its transformation into meat induces a series of biochemical and physiological changes: pH falls, ionic strength varies, and proteolysis and protein interactions (Myosin/Actin) occur. These alterations create physical modifications. Membrane functionality is altered, as is its physical integrity. Processes such as cooking or salting further destroy these structures. In these conditions, histological pictures which associate water fractions determined by NMR with intra- and extra-cellular domains, as previously proposed, remain unlikely [15].
MRI provides morphological images that can be associated with parametric images of relaxation times, or diffusion within the tissue at different post-mortem time-points. Based on T_2 maps hypersignals were able to highlight free water accumulation and the formation of channels containing this mobile water in close proximity to the connective tissue[16,17].. These results were confirmed in water diffusion studies which demonstrate that water in drip channel diffuses more freely and isotropically[18]. The influence of intracellular diffusional barriers has been investigated[19]. The anisotropy of diffusion coefficient in meat was shown in the early post mortem period, and its decrease with the post mortem time could be related to the membrane degradation[20]. As for relaxation, diffusion decays can be acquired to resolve possible water compartments differing from their diffusion characteristics. In meat, bi-exponential behaviour of diffusion decay was observed which persists post mortem during ageing (see figure 2). Presumably because of partial volume effect, a biexponential behaviour was detected in region showing a free water reservoir. These results underline the usefulness of diffusion tensor measurements for characterizing muscle structure, and help to understand the mechanisms of post mortem water exudation[21].
T_2 maps of a muscle sample slice 6 hours post mortem. Diffusion-weighted signal decay curves from a voxel in the myofiber region and the region of the water reservoir measured in a direction perpendicular to the main fiber direction. The biexponential model is given as a solid line.

2.3 PROCESSING

Meat is subjected to various technological processes before consumption, of which slaughter, storage, freezing, thawing, salting and cooking are the most common. The aim of this paper is not to review MRI applications for each process. We focused our presentation on curing, which is one of the oldest technologies for food preservation. Salt (sodium chloride) is added as a flavouring or flavour enhancer, as a preservative, and as a way of drawing out desirable textural characteristics in meat products. However, excessive salt consumption is linked to cardiovascular disease, and the health authorities recommend reducing the salt content of brined products. To achieve a safe salt reduction, it is necessary to define the best curing conditions. How can these conditions be achieved and checked? Analyses can be performed on the salting process to determine 1) the distribution of salt, by ^{23}Na NMR, 2) the density and mobility of water and apparent diffusion coefficients, by ^{23}Na and ^{1}H NMR, and 3) the ratio of bound to free Na^+ ions. Thus, NMR

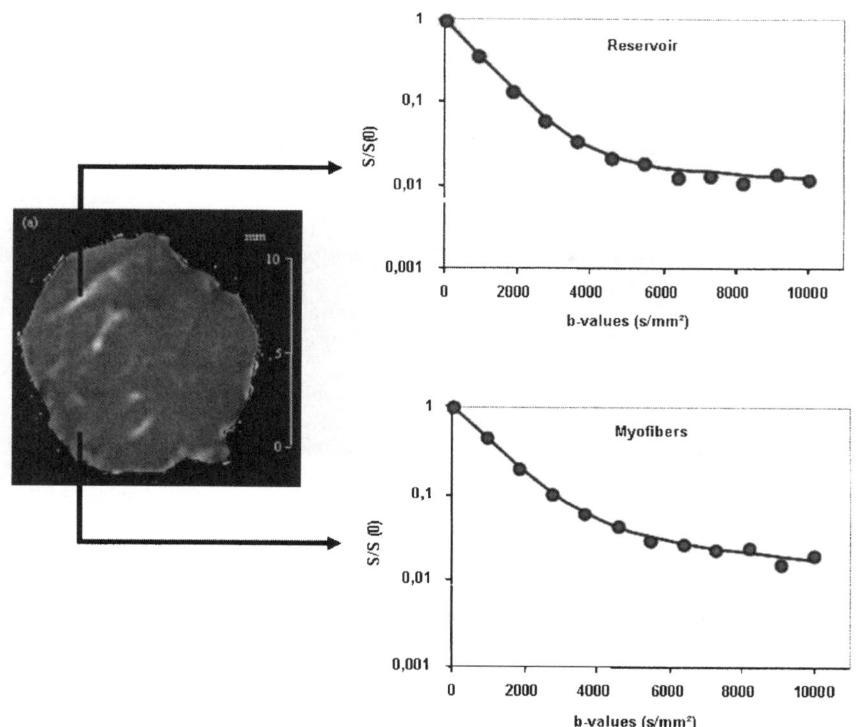

Figure 2 *T_2 maps of a muscle sample slice 6 hours post mortem; Diffusion-weighted signal decay curves from a voxel in the myofiber region and the region of the water reservoir measured in a direction perpendicular to the main fibber direction. The biexponential model is given as a solid line*

offers a unique opportunity to non-invasively assess the distribution and status of Na^+ and Cl^- ions in tissues [22 23 24].

2.3.1 Distribution of salt

The ^{23}Na MRI was used to study the distribution of brine in muscle according to the salting process[25]. This ^{23}Na NMR study clearly showed that salt distribution was heterogeneous within the ham and was dependent on technological process, underling that MRI is a suitable tool. These ^{23}Na NMR images can be compared with the images generated by ^{1}H NMR. MRI allows a simultaneous visualisation of the distribution of Na^+ and the distribution of different tissues (lean, bone, fat) in a heterogeneous product such as meat.

The effect of the fat content on the heterogeneity of salt distribution was investigated in smoked fillet of Atlantic salmon [26]. Figure 3 shows transverse images of for salmon fillets (skin side down). The fillet samples were taken from either the anterior or posterior parts and had undergone different salt processing times. Three reference tubes with different known salt concentration were placed on the top of fillets. Sodium distribution varied between fillets. The highest concentration (hypersignal) was found in the thickest fillet (posterior part) shown in Figure 3a. Sodium distribution within fillets was heterogeneous. Oblique dark runs (hyposignal) were detected, especially in fillet 1 (anterior part salted 5 h). These areas corresponded to myocommata which separates muscle segments. Fats were found within the connective tissue in these myocommata. Fat distribution is clearly visible in Figure 3b showing a water-suppressed image. Moreover, salt concentrations were lower in the skin side of the fillet than in the belly side, as shown on each fillet bottom.

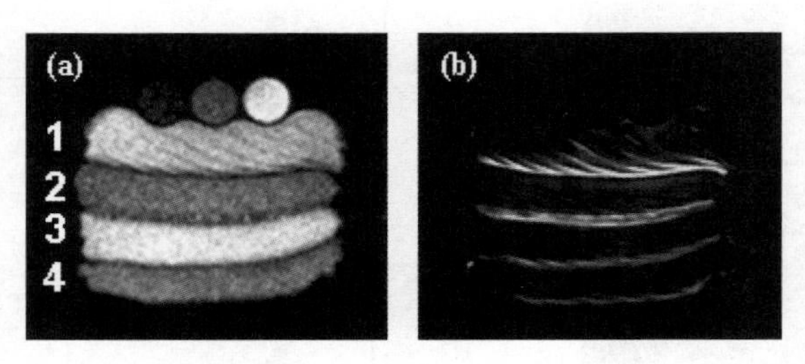

Figure 3 *^{23}Na (a) , and water suppressed (b) MR-images of salted-smoked Atlantic salmon fillets. From top to bottom; lateral sections (skin side down) of the anterior part salted 5 h, anterior part salted 1.5 h,the posterior part salted 5 h and the posterior part salted 1.5 h. Reference tubes (on the top) contained NaCl-doped water solutions of 1, 2.5, and 5% from the left to the right hand side, respectively.*

Using reference tubes, salt content was determined in each voxel as the water T_2. The results highlighted that the highest salt concentrations in the fillet were associated with the lowest T_2 values (Table 1). The salt therefore induces a decrease in water mobility.

Fillet n°	Salt content (%)	T_2 (ms)
1	4.2 ± 0.3	35 ± 2
2	3.0 ± 0.2	39 ± 2
3	4.9 ± 0.3	34 ± 1
4	3.0 ± 0.2	38 ± 2

Table 1 *Salt content and mean T_2 values extracted from the ^{23}Na and T_2 images*

2.3.2 Dynamic study of salting process

Na^+ diffusion was measured to investigate the changes induced by sodium diffusion in gelatin gel during salting. The ^{23}Na MR images (Figure 4) provided clear evidence of progressive sodium diffusion into the gel matrix increasing over time [27]. The salt decrease in brine and the salt increase in gelatin were both exponential processes, while total ^{23}Na signal intensity (brine+gelatin) did not show any significant variation, suggesting that nearly all the salt content was revealed by ^{23}Na MR imaging (Figure 5). Furthermore, time resolution was compatible with the rate of sodium diffusion.

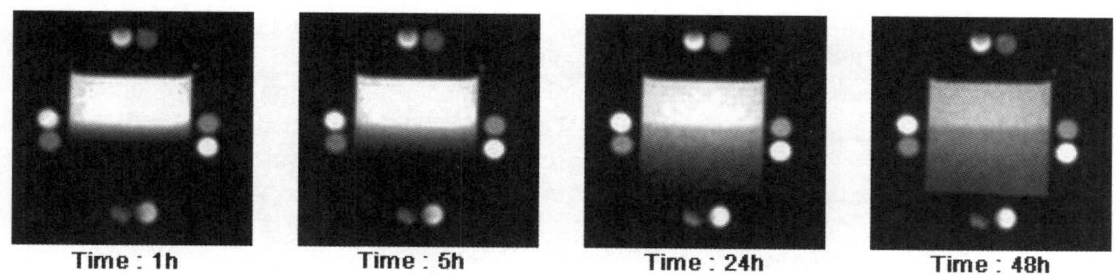

Figure 4 *^{23}Na MRI of 1M NaCl brine (top) and 25% gelatine gel (bottom) at different contact times.*

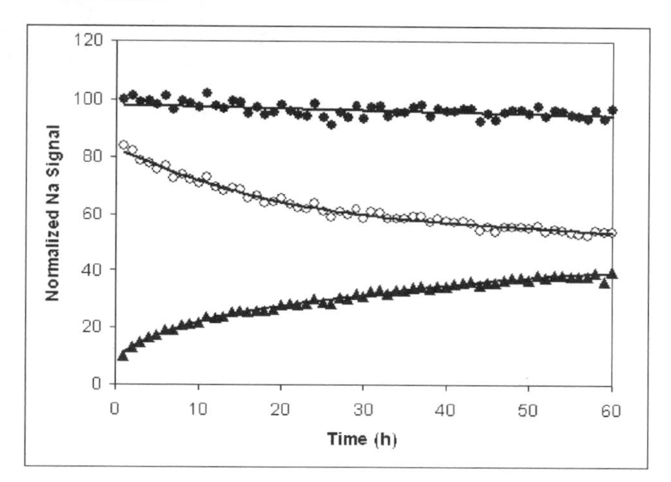

Figure 5 *Evoloution of Na^+ concentration (o) in brine, (▲) in gel and (●) total (salt +brine)*

The Na^+ profile according to the diffusion direction gave more information about the spatial heterogeneity (Figure 6). After 72 hours in gelatin (homogeneous product), Na^+ concentrations were homogeneous in the first 7 mm, decreasing slowly through the next 13 mm before dropping to the bottom of the gelatin.

The ingress of sodium and salt uptake in a piece of cod (G*adus morhua*) fillet was also studied by real-time MRI, and showed a heterogeneous distribution of sodium, presumably due to local differences in the matrix microstructure or diffusion distances[28]. In the same way, a study on movements of sodium ions in pork loin during brining was carried out using ^{23}Na MRI [29]. The ^{23}Na profiles obtained over 5 days at 24-hr intervals suggested that the diffusion of salt into whole meat cuts cannot be described by simple ordinary Fickian diffusion with a constant diffusion coefficient. It has been suggested that the diffusion coefficient is affected by changes in NaCl concentration, swelling, and degree of dehydration.

^{1}H and ^{23}Na NMR approaches make it possible to establish a correlation between water fluxes and Na^+ ingress in meat during salting. They should therefore make it possible to optimize processing methods according to the raw products employed (freshness, fat distribution, etc.)[30 31].

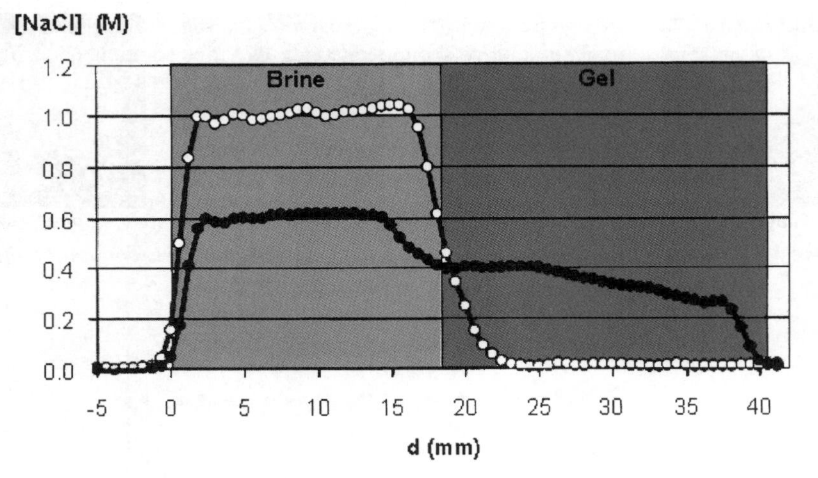

Figure 6 *^{23}Na concentration along the direction of salt diffusion (perpendicular to the front brine / gel at two times 1 (o) and 72 (•) hours after contact.*

2.3.3 *Ratio (Na^+free /Na^+ bound)*

^{23}Na has a 3/2 spin and therefore three single quantum transitions. For free ions, these transitions have the same resonance frequency under conditions of extreme narrowing ($\omega_0 \tau_c << 1$). In interaction with macromolecules, Na^+ ions undergo static and dynamic quadrupolar effects. Accordingly, the -3/2 ⇔ -1/2 and 1/2 ⇔ 3/2 transitions decay at a faster rate than the-1/2 ⇔ 1/2 transition. This quadrupolar property can be used to measure the ratio of free Na^+ to bound Na^+ ions *in situ*[32 33] A Double Quantum Filter (DQF) will select only the NMR signal belonging to the ions interacting with macromolecules. To the evaluate DQF signal accurately, there has to be an optimization of the double quantum creation time τ in the pulse sequence $\pi/2 - \tau/2 - \pi - \tau/2 - \pi/2 - \delta - \pi/2 -$ Acq(t). This can be achieved by running a set of experiments at different τ values and fitting the function $[\exp(-\tau/T_{2S}) - \exp(-\tau/T_{2F})]$ to the peak heights of the resultant spectra, as illustrated in Figure 7 for fresh pork meat and dry sausage.

DQF signal intensity increases for very short τ (characterized by the time constant T_{2F}), then goes through a peak (optimal τ value; τ_{opt}), before finally decaying with T_{2S}. The data were fit to the function mentioned above, which gave the determination of τ_{opt}: 1.6 and 10 ms for dry sausage and fresh pork meat, respectively.

It has previously been shown that both T_{2F} and T_{2S} relaxation times are tissue-dependent[34]. DQF signal intensity depends on the strength of the quadrupolar interaction of sodium ion with its environment. Macromolecules of different sizes and dynamics are expected to exhibit different rise and decay times.

There have been a few reports on multiple filtration quantum sodium imaging that have provided images of bound sodium ions difficult to detect in conventional experiments[35, 36]. Figure 8 shows single-quantum and DQ-filtered gradient-echo images of a 5 mm NMR tube inside and coaxial with a 10 mm tube. The inner tube was filled with saline solution (1.5 M), the outer annulus with sausage (10% of salt w/w). The DQ-filtered image clearly demonstrates the position of the sodium-containing sausage with excellent signal suppression from isotropically mobile sodium nuclei of the saline solution.

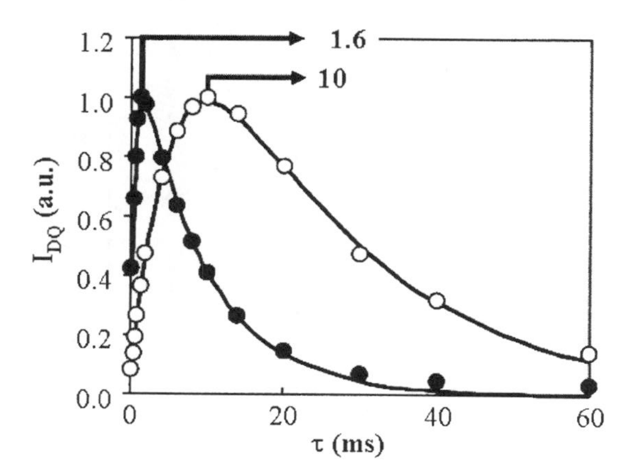

Figure 7 *DQ intensity (I_{DQ}) as a function of the creation time τ, for (o) fresh meat and (•) dry sausage obtained at 9.4 T (105.8 MHz for ^{23}Na)*

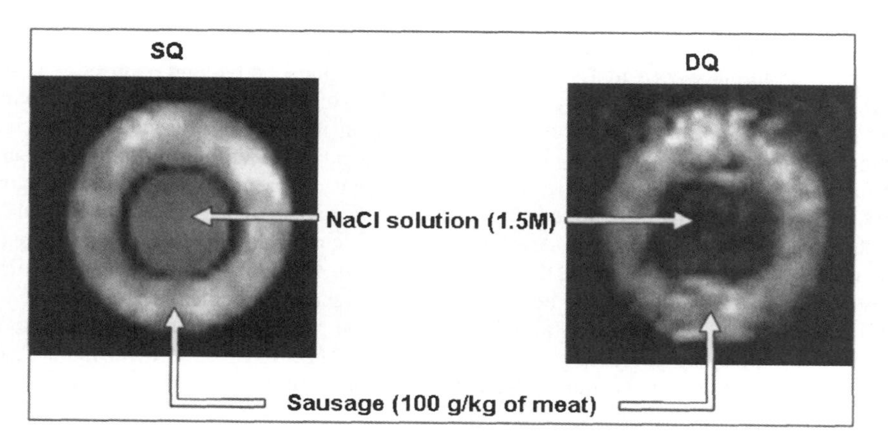

Figure 8 *A single quantum image (SQ) and a double quantum (DQ) image of sausage sample with a reference brine tube.*

An SQ/DQ study on pork meat prepared at different sodium concentrations (Figure 9) demonstrated the excellent correlation between $[Na]_{NMR}$ and $[Na]_{added}$ (all visible sodium)[33]. The decrease in bound Na^+ fraction, as determined from the DQ experiment, with the increase in salt content is consistent with a saturation of the number of fixing sites.

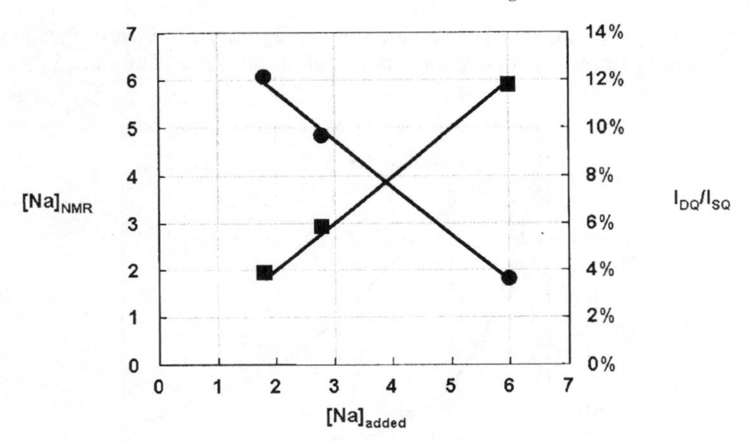

Figure 9 *Sodium content determined by single-quantum NMR ($[Na]_{NMR}$) and ratio between double and single-quantum NMR signals (I_{DQ}/I_{SQ}) vs. the salt added to pork meat ($[Na]_{added}$). The salt contents are expressed in % (w/w).*

3 CONCLUSION

MRI is a useful tool in food research. It gives a highly accurate determination of meat composition in pieces of carcass such as total fat, subcutaneous fat, and connective tissue, and enables muscle architecture to be shown in 3D. This characterization of raw meat is can provide deeper insight into the mechanisms responsible for sensory properties such as tenderness and palatability. This knowledge of raw products is doubly important in that it enables technological processes to be adapted to meat characteristics. The technological process could be further improved direct monitoring in the magnet. However, all these studies require methodological developments, and certain highly sophisticated pulse sequences such as DQF and TDI need to be run in order to obtain the information. Research in MRI methods designed for medical diagnostic imaging is generating very useful results, but the devices required to monitor the process in the magnet have to be adapted to mimic the processes undergone by meat. These devices have to induce zero disturbance in NMR signal acquisition and be able to work in strong magnetic fields.

References

1. P.A. Fowler, M.F. Fuller, C.A. Glasbey, G.G. Cameron, and M.A. Foster, *Am. J. Clin. Nutr.,* 92, **56**, 7.
2. A. Davenel, Seigneurin, F., Collewet, G., and Remignon, H., *Meat Science,* 2000, **56**, 153.
3. G. Collewet, P. Bogner, P. Allen, H. Busk, A. Dobrowolski, E. Olsen, and A. Davenel, *Meat Science,* 2005, **70**, 563.
4. M. Monziols, G. Collewet, M. Bonneau, F. Mariette, A. Davenel, and M. Kouba, *Meat Science,* 2006, **72**, 146.
5. M. Monziols, G. Collewet, F. Mariette, M. Kouba, and A. Davenel, *Magnetic Resonance Imaging,* 2005, **23**, 745.
6. Offer, G. and Knight, P. The structural basis of water-holding in meat. Part 1: general

principles and water uptake in meat processing. In Lawrie, R.A. (ed.) Development in meat science. Elsevier Science Publishers, London (1988).
7. Offer, G. and Knight, P. The structural basis of water-holding in meat. Part 2: Drip losses. In Lawrie, R.A. (ed.) Development in meat science. Elsevier Science Publishers, London (1988).
8. J.P. Renou, J. Kopp, P. Gatellier, G. Monin, and G. Kozak-Reiss, *Meat Science.,* 89, **26**, 101.
9. J. Brøndum, L. Munck, P. Henckel, A.H. Karlsson, E. Tornberg, and S.B. Engelsen, *Meat Science,* 2000, **55**, 177.
10. Borowiak P. , J. Adamski, K. Olszewski, and J. Bucko. In: 32th European meeting of meat research workers, Ghent, 1986
11. R.J. Brown, F. Capozzi, C. Cavani, M.A. Cremonini, M. Petracci, and G. Placucci, *J. Magn. Reson.,* 2000, **147**, 89.
12. S. Fjelkner-Modig and Tornberg, E., *Meat Science,* 1986, **17**, 213.
13. Tornberg, E. , Andersson, A., Göransson, A., and Von Seth, G. Water and fat distribution in pork in relation to sensory properties. In Puolanne, E. and Demeyer, D. (eds.) Pork quality, genetic and metabolic factors. CAB International, Townbridge (1993).
14. J.P. Renou, G. Monin, and P. Sellier, *Meat Science.,* 85, **15**, 225.
15. A. Traoré, Foucat, L., and Renou J.P, *Biopolymers,* 2000, **53** , 476.
16. J.-M. Bonny, W. Laurent, and J.P. Renou. In: 7th ISMRM, Philadelphie, 1999
17. H.C. Bertram , A.K. Whittaker, H.J. Andersen, and A.H. Karlsson, *Meat Science,* 2004, **68**, 667.
18. G. Offer and T. Cousins, *J. Sci. Food. Agric.,* 92, **58**, 107.
19. S.T. Kinsey, Locke, B.R., Penke, B., and Moerland, T.S., *NMR in Biomedicine,* 1999, **12**, 1.
20. L. Foucat, Benderbous, S., Bielicki, G., Zanca, M., and Renou, J.P., *Magnetic Resonance Imaging,* 1995, **13**, 259 .
21. J.M. Bonny and Renou, J.P., *Magnetic Resonance Imaging,* 2002, **20**, 395.
22. G. Navon, H. Shinar, U. Eliav, and Y. Seo, *NMR in Biomedicine,* 2001, **14**, 112.
23. P.S. Belton, Packer, K.J., and Southon, T.E., *J. Sci. Food. Agric.,* 1987, **41**, 267.
24. T. Nagata, Chuda, Y., Xiaojun Yan, M., Suzuki, M., and Kawasaki, K.-I., *J. Sci. Food. Agric.,* 2000, **80**, 1151.
25. J.P. Renou, Benderbous, S., Bielicki, G., Foucat, L., and Donnat, J.P., *Magnetic Resonance Imaging,* 1994, **12**, 131 .
26. L. Foucat, R. Ofstad, J.-P. Donnat, and J.-P. Renou. In: 7th International Conference on Applications of Magnetic Resonance in Food Science, Copenhagen - Denmark , 2004
27. A. Traoré , M. Mouaddab, J.-P. Donnat, L. Foucat, and J.-P. Renou. In: IUFOST XIIIth World Congress of Food Science and Technology FOOD IS LIFE, Nantes (France), 2006
28. U. Erikson, Veliyulin, E., Singstad, T.E., and Aursand, M., *J Food Sci ,* 2004, **69**, 107.
29. C. Vestergaard, J. Risum, and J. Adler-Nissen, *Meat Science,* 2005, **69**, 663.
30. E. Veliyulin, I. Aursand, and U. Erikson. In: 8th International Conference on the Applications of Magnetic Resonance in Food Science, University of Nottingham, UK, 2006
31. C.L. Hansen, F. van den Berg, S. Ringgaard, A.H. Anders H. Karlsson, and Střdkilde-Jřrgensen H. In: 8th International Conference on the Applications of Magnetic Resonance in Food Science, University of Nottingham, UK, 2006
32. Foucat, L., Donnat J.P. , and Renou J.P. ^{23}Na and ^{35}Cl NMR studies of the interactions

of sodium and chloride ions with meat products. In Belton, P.S., Gil, A.M., Webb, G.A., and Rutledge, D. (eds.) Magnetic Resonance in Food Science latest developments. Royal Society of Chemistry, Cambridge (2003).

33. M. Mouaddab, L. Foucat, J.-M. Bonny, and J.-P. Renou. In: 8th International Conference on the Applications of Magnetic Resonance in Food Science, University of Nottingham, UK, 2006
34. A. Neufeld, U. Eliav, and G. Navon, *Magn Reson Med,* 2003, **50**, 229.
35. M.D. Cockman , Jelinski, L.W., Katz, J., Sorce, D.J., Boxt, L.M., and Cannon, P.J., *J. Magn. Reson.,* 1990, **90**, 9.
36. S. Wimperis, Cole, P., and Styles, P., *J. Magn. Reson..* 1992, **98**, 628.

USE OF MRI TO PROBE THE WATER PROTON MOBILITY IN SOY AND WHEAT BREADS

A. Lodi and Y. Vodovotz

Department of Food Science and Technology, The Ohio State University, Parker Food Science and Technology Bldg., 2015 Fyffe Rd., Columbus, OH 43210

1 INTRODUCTION

Various factors have been shown to play a critical role in staling of baked goods including amylopectin recrystalization, changes in the amorphous domains and interactions among other bread components (e.g.: proteins and lipids). A key factor in these changes is the distribution and mobility of water. For example, a more uniform water distribution in fresh white bread has been shown to decrease the staling rate[1] and may affect the amylopectin retrogradation as well since this process requires binding of water molecules.[2] States and distribution of water in bread have been locally investigated using techniques such as nuclear magnetic resonance (NMR), differential scanning calorimetry (DSC), thermogravimetric analysis (TGA). However, the majority of these techniques requires very small sample size and therefore relate to specific portions of the bread loaf (usually the central part of the bread crumb is investigated). The intrinsic inhomogeneity of bread samples dictates the use of experimental techniques able to investigate the largest possible samples, so that a more accurate representation is observed.

Magnetic resonance imaging offers a noninvasive method for measuring water mobility and distribution of the entire bread slice simultaneously.

Inclusion of soy ingredients in bread formula has been considered of interest for many years as a high quality protein enrichment. More recently, soy ingredients have also been shown to possess health promoting activity, due to an association between consumption of soy protein, isoflavones and other phytochemicals (saponins, lignans, and phytic acids) and a reduced risk of developing chronic diseases.[3,4] Moreover, the addition of soy ingredients to bread formulations has been shown to hinder the staling rate.[1] However, the major limitation to the amount of soy that can be added to bread is the dramatic impact it has on loaf quality (specific loaf volume in particular).[5-7] Recently, a highly acceptable soy bread containing enough soy protein to meet the FDA health claim[8] has been developed in our laboratory.[9] The addition of soy ingredients to bread required inclusions of larger amounts of water in the formulation (as previously reported by Ofelt[10,11]), due to the different hygroscopic nature of soy. This can be hypothesized to affect the states of water molecules in the bread matrix of the fresh and stored bread that may lead to product stability and affect staling rate. Therefore, the objective of this study was to characterize water distribution and mobility in soy and wheat breads using MRI.

2 METHOD AND RESULTS

The soy containing bread samples were formulated and produced according to patent pending procedures.[9] The wheat bread samples were prepared in a bread machine. After baking the loaves were allowed to cool to room temperature and sealed in polyethylene bags. Bread was stored in these bags at 4 °C for the duration of the study. No further sample preparation was required, due to the non destructive and non invasive characteristics of MRI.

The MRI experiments were performed using a 4.7 T / 40 cm magnet controlled by a Bruker Avance Console. A 260 mm inner diameter gradient coil with a 200 mm inner diameter proton volume RF coil was used. A spin echo pulse sequence was performed in all the experiments. 8 images were acquired using 15.37 ms echo time and 2 s repetition time for the soy bread samples. For the wheat bread sample, which was found to have shorter T_2 value than the soy based breads, 8 images with shorter echo time (11.87 ms) and 2 s repetition time were also acquired. 5 slices, 10 mm thick, were excited perpendicular to the longest dimension of the loaves using a FOV (256 × 256 matrix) of 150 × 150 mm^2 for the soy-containing samples, and 165 × 165 mm^2 for the wheat bread.

Relaxation times relate to molecular rotational diffusion, therefore changes of T_2 values provide a good indication of changes in water protons mobility occurring in the analyzed samples during storage. The first of the 8 images acquired for the calculation of T_2 maps was used as a proton signal intensity map to investigate water distribution throughout the loaf section. During data acquisition, each loaf was inserted in the magnet next to a H_2O/D_2O phantom (that contained enough water to have signal intensity comparable to the bread samples). The phantom signal was used to normalize the intensity images from different acquisitions.

Images were acquired on fresh samples and on days 1, 3, and 6 of storage to investigate distribution and mobility of water protons across sections of the bread loaves. Loaves were removed from the magnet between experiments and thus no point-by-point comparison was possible.

2.1 Water proton distribution

Water proton intensity magnetic resonance images of fresh wheat and soy bread samples are presented in Figure 1. The lower water content in the wheat sample (about 39%, wet basis), as compared to the soy bread (about 44%), resulted in a signal of lower intensity.

The distribution of the signal was more homogeneously distributed throughout the slice section in the soy bread. This was attributed to the higher amount of water introduced in the soy bread formula and to the stronger water binding occurring in presence of soy ingredients. Moreover, the soy bread had lower specific loaf volume as compared to the wheat sample,[5-7] which may have contributed to a decrease in water evaporation during baking. Homogeneity of water distribution in bread samples has been previously shown to hinder the staling rate[1] and therefore, freshness retention may be greater in the soy containing loaves.

During storage little change was observed for the water proton distribution in the soy bread (Figure 2). On the other hand, water proton signal on the sixth day of the wheat bread (Figure 2) exhibited an average lower intensity than the fresh bread (Figure 1) and was more evenly distributed throughout the slice section (in particular towards the bottom portion of the loaf). In wheat bread a stronger driving force to water migration was present (from crumb to crust), which caused relevant water migration and distribution of the signal throughout the slice section in the stored sample.

Therefore, homogeneity of water distribution and limited changes of water states during the considered storage period, observed in whole loaves of soy bread, may improve storage stability as compared to the wheat bread.

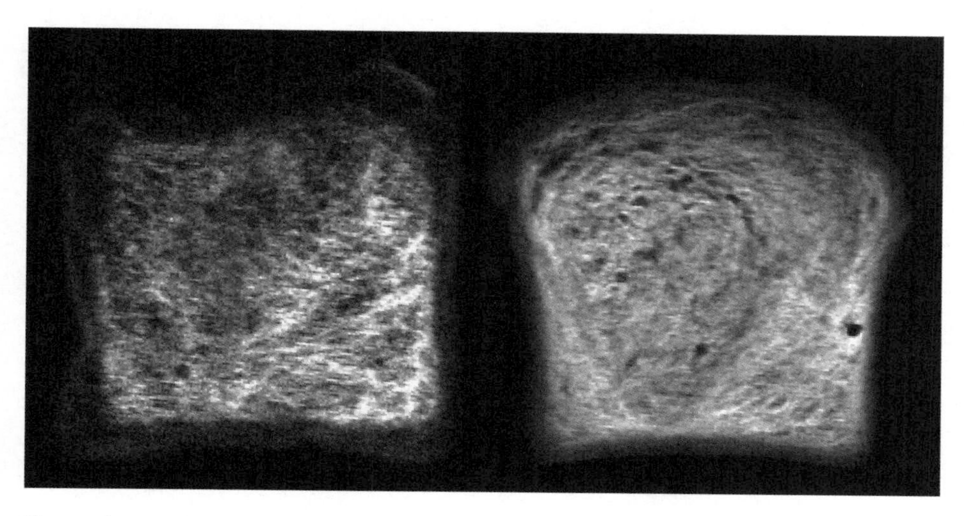

Figure 1 *Proton intensity signal images of fresh (day 0 of storage) wheat (left) and soy (right) bread samples*

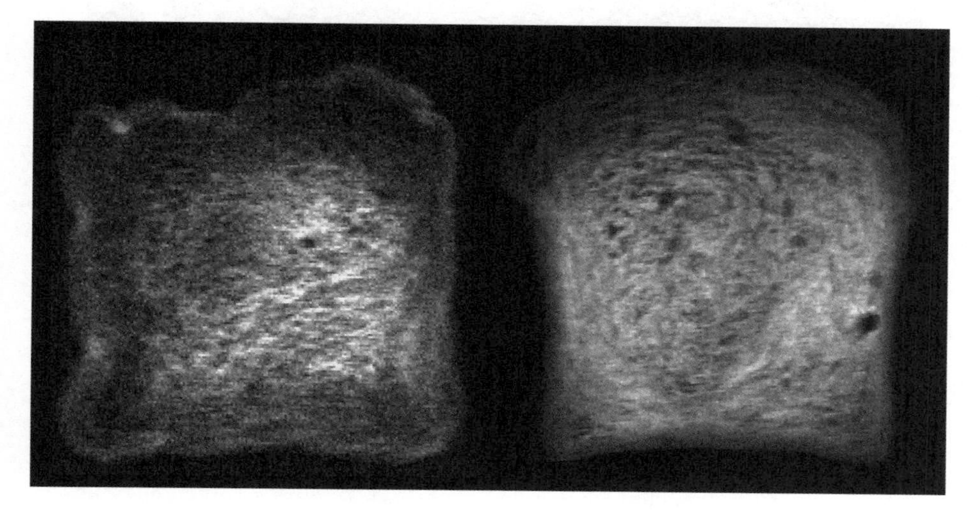

Figure 2 *Proton intensity signal images of wheat (left) and soy (right) bread samples at day 6 of storage*

2.2 Water proton mobility

Changes of water protons mobility (as probed by the spin-spin relaxation time, T_2) in wheat and soy breads were also investigated and reported in Figure 3 (fresh wheat and soy breads) and Figure 4 (stored samples).

The homogeneous appearance of the intensity maps observed for the soy bread products was also observed for T_2 images. The water protons T_2 values for the fresh soy-containing sample (Figure 3) were found to be distributed around a value of ~18 ms. The wheat bread sample exhibited a lower T_2 value (~10 ms) and, as seen in the intensity images, the signal concentrated in the central part of the loaf. The outmost part of the wheat bread loaf, corresponding to the crust and a layer underneath it, generated a very low signal (too solid-like). During storage, distribution of T_2 values for soy bread had minor changes (remained homogeneous), however the average T_2 value of water protons decreased to about 15 ms at day 6 of storage (Figure 4). The distribution of T_2 values across the slice of fresh (Figure 3) and stored (Figure 4) wheat bread was found to vary significantly. In the wheat bread sample this was likely due to the combined effect of water protons migration (as seen in Figure 1 and 2) and decrease of T_2 (the average T_2 value was calculated to be around 4 ms when considering the same number of non-zero points as in the day 0 slice). The decrease of T_2 values in the bread products, observed previously by other authors,[12] has been attributed to a progressive strengthening of water binding to macromolecules in bread during storage. Overall, the larger decrease of T_2 values in wheat bread may indicate that the soy-containing product can better retain freshness. The limited changes observed for the soy containing products, partly due to the strong homogeneity of the fresh product and to a smaller driving force for moisture migration, may relate to a hindered staling rate.

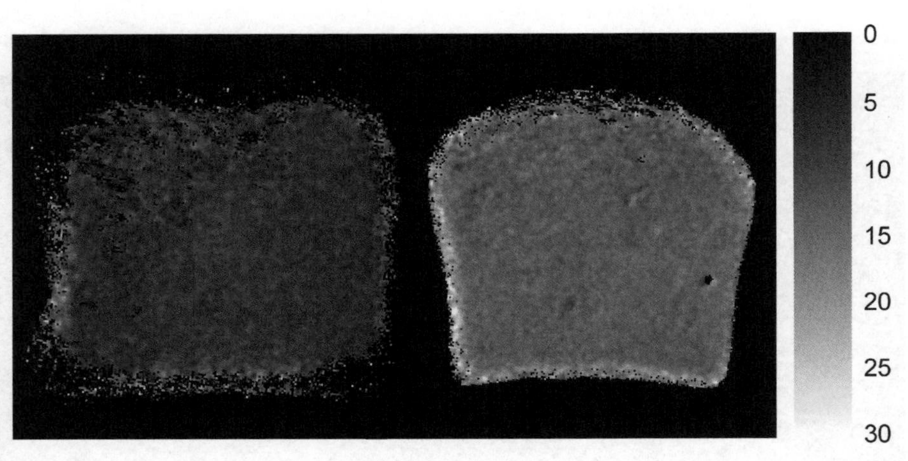

Figure 3 *T2 images of fresh (day 0 of storage) wheat (left) and soy (right) bread samples*

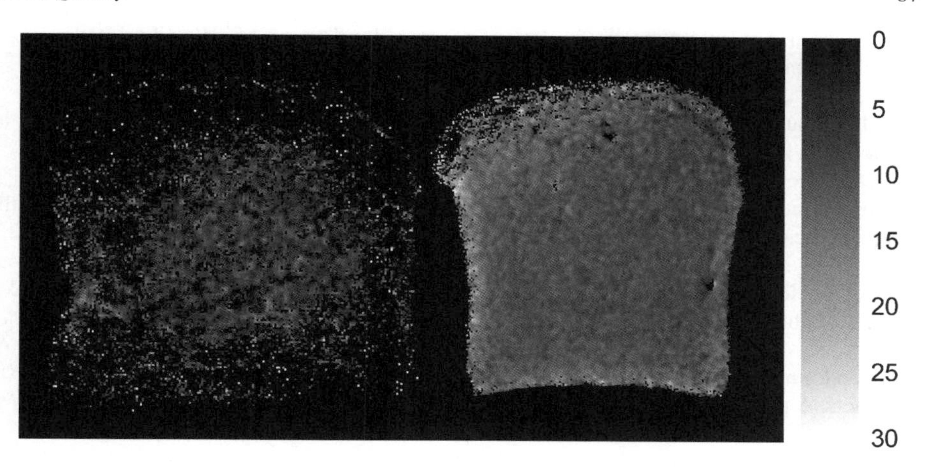

Figure 4 *T_2 images of wheat (left) and soy (right) bread samples at day 6 of storage*

3 CONCLUSIONS

Changes in distribution and dynamics of water in bread with and without soy were investigated using magnetic resonance imaging. The addition of soy was found to improve the homogeneity of water distribution in freshly baked breads, which prevents water migration and may hinder the rate of staling processes.

Mobility of water protons, as probed using spin-spin relaxation time, was found to undergo a smaller decrease during 6 days of storage in bread samples containing soy than in traditional wheat bread. Slower changes in water dynamics may also induce slower staling processes. Therefore, the addition of relevant amounts of soy may improve freshness retention in bread.

Magnetic resonance imaging provides a powerful tool to investigate changes and dynamics of single components in complicated food systems in a non destructive and non invasive fashion.

REFERENCES

1. Maga, J. A., Bread staling. *CRC Critical Reviews in Food Technology* 1975; **5**: 443-492.

2. Gray, J. A.; Bemiller, J. N., Bread staling: molecular basis and control. 2003; **2**: 1-21.

3. Erdman, J. W., Jr., AHA Science Advisory: Soy protein and cardiovascular disease: A statement for healthcare professionals from the Nutrition Committee of the AHA. 2000; **102**: 2555-9.

4. Potter, S. M., Soy protein and cardiovascular disease: the impact of bioactive components in soy. 1998; **56**: 231-5.

5. Mizrahi, S.; Zimmermann, G.; Berk, Z.; Cogan, U., The use of isolated soybean proteins in bread. *Cereal Chem.* 1967; **44**: 193.

6. Pomeranz, Y.; Shogren, M. D.; Finney, K. F., Improving breadmaking properties with glycolipids. I. Improving soy products with sucroesters. *Cereal Chem.* 1969; **46**: 503.

7. Pomeranz, Y.; Shogren, M. D.; Finney, K. F., Improving breadmaking properties with glycolipids. II. Improving various protein-enriched products. *Cereal Chem.* 1969; **46**: 512.

8. FDA, Food labeling: health claims; soy protein and coronary heart disease. Food and Drug Administration, HHS. Final rule. 1999; **64**: 57700.

9. Vodovotz, Y.; Ballard, C. Formula and Process for Making Soy-Based Bakery Products. A patent application, Docket No. 10-267845. 2002.

10. Ofelt, C. W.; Smith, A. K.; Derges, R. E., Baking behavior and oxidation requirements of soy flour. I. Commercial full-fat soy flours. *Cereal Chem.* 1954; **31**: 15.

11. Ofelt, C. W.; Smith, A. K.; Mills, J. M., Baking behavior and oxidation requirements of soy flour. II. Commercial defatted soy flours. *Cereal Chem.* 1954; **31**: 23.

12. Chen, P. L.; Long, Z.; Ruan, R.; Labuza, T. P., Nuclear magnetic resonance studies of water mobility in bread during storage. *Food Sci. Technol.-Lebensm.-Wiss. Technol.* 1997; **30**: 178.

PROBING WATER MIGRATION AND MOBILITY DURING THE AGEING OF BREAD

N.M. Sereno[1], S.E. Hill[1], J.R. Mitchell[1], U. Scharf[2] and I.A. Farhat[1]

[1]Division of Food Sciences, School of Biosciences, University of Nottingham, Sutton Bonington Campus, Loughborough, LE12 5RG, UK
[2] Firmenich S.A., CH-1217 Meyrin, Switzerland

1 INTRODUCTION

The wide spatial variation in water contents throughout freshly baked baguette-style breads gives the product its desired eating quality with a brittle glassy crust and a soft rubbery crumb. The loss of quality during storage results from the concurrent changes to structure and composition leading to changes to texture and flavour. This phenomenon is referred to as bread staling, and from a texture viewpoint, it is described as a general decrease in the crispness of the crust and an increase in the firmness of the crumb coupled with a dryer mouth feel.

The 2 main mechanisms responsible for these changes (often occur concurrently) are:

Starch retrogradation: this refers to the reordering of amylose and amylopectin post-baking. This reordering can proceed to a recrystallisation detectable by wide-angle x-ray diffraction. The effects of storage temperature, water content and amylose on retrogradation have been studied extensively. Other related phenomena such as changes to the gluten protein network, the interactions between gluten and starch, and the influence of amylose-lipid complex are usually invoked in bread staling studies [1-13].

Water migration: this covers 3 different aspects:

a. The loss of water from the bread.
b. The spatial redistribution of water between crumb and crust.
c. The repartitioning of water between the various components (starch *versus* gluten, amorphous *versus* crystalline starch, etc.) of the bread at the molecular level.

This study addresses starch retrogradation and the macroscopic water migration processes (e.g. a. and b.).

2 MATERIAL AND METHODS

Freshly baked commercial small French-style baguettes (~100g each) were studied. The overall water content of the baguettes was ~30% wet weight basis (wwb) as defined by drying at 105°C.

2.1 Gravimetric study of water migration

The baguettes were stored in a controlled environment (20°C and RH=54%) to mimic usual domestic conditions. In order to monitor the overall change in water content as well as the distribution between crumb and crust, the following water contents were determined at regular intervals (from the weight changes):

- The total water content.
- The water contents of the outer third (by weight) of the baguette (labelled as "the crust"), the inner third ("the inner crumb") and the intermediate fraction ("the outer crumb"). It is worth noting that "the crust" refers to the outer 1/3 of the baguette and not only to the glassy fraction traditionally known as crust, the thickness of which can vary making any meaningful mass balance calculations difficult.

2.2 Nuclear Magnetic Resonance (NMR)

A low field spectrometer (MARAN, Oxford Instruments) operating at 23 MHz and 20°C was used to acquire Free Induction Decays (FID) using the standard single 90° pulse or the solid-echo sequences and the Carr-Purcell-Meiboom-Gill (CPMG) spin-echo decay.

Two types of samples were analysed:

- Samples of each part (inner crumb, outer crumb and crust) of baguettes collected from the ageing experiment described above (20°C and RH=54%). This allowed the study of the combined effects of starch retrogradation and water migration, which occur during bread storage.
- In order to study the relative contribution of (i) the decrease in water content and (ii) the re-ordering and subsequent crystallisation of starch (i.e. starch retrogradation), to the changes in molecular mobility during ageing, two sets of experiments were performed:
  - Study the impact of starch retrogradation on molecular mobility at constant water content. Samples of fresh baguette outer crumb (~41% water) were sealed in the NMR tube and monitored at regular intervals over a period of 7 days.
  - Study the effect of water content on a fully retrograded sample (outer crumb aged for 98h at 20°C at a constant water content of ~41% water), which was re-equilibrated to a range of water contents.

3 RESULTS AND DISCUSSION

3.1 Water migration during ageing

The complete baguette had an initial water content of ~30% wwb. As was expected, during storage at 20°C and RH=54%, the water content decreased gradually until it reached 13% wwb after 7 days.

Figure 1 illustrates how the water was distributed throughout the baguette in the fresh baguettes and the changes during ageing. Initially the crumb contained most (~83%) of the present water, but after 7 days of storage, and following extensive loss and redistribution of water, the remaining water was equally distributed along the cross section of the baguette. When considering the baguette components individually a steady decrease in water levels was observed for the inner and outer crumb. The water content in the crust however increased slightly during the first two days, suggesting that the water intake by the crust, due to the migration of water from the crumb to the surface, was greater than the loss from the crust to the surrounding air. Such crumb/crust water migration promoting a reduction in crispness by the crust is often suggested but rarely quantified [10].

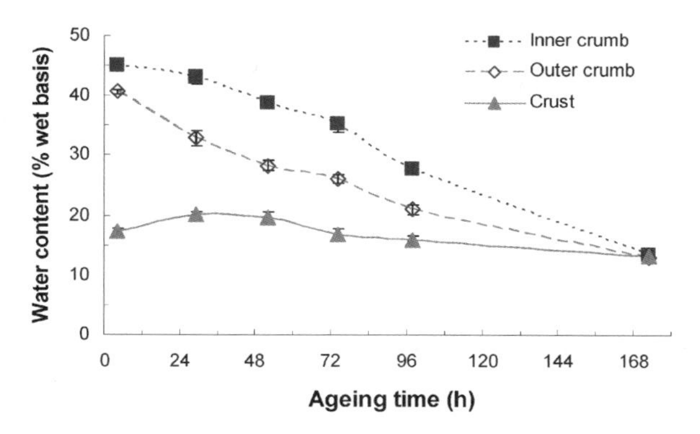

Figure 1 *The changes in water content of the 3 fractions of a baguette during ageing. The error bars correspond to 1 standard deviation of triplicate measurements.*

3.1 Changes in molecular mobility during ageing

Bread staling is associated with significant increase in texture hardness and mouth feel dryness. Time domain NMR has often been used to monitor changes occurring during starch retrogradation at constant water contents [4,6,11-13]. In general, an overall decrease of molecular mobility is observed. However, to our knowledge, studies deconvoluting the roles of changes to total and local water contents from those due to retrogradation have not been carried out on bread in a systematic way.

- A typical example of the changes in the proton Free Induction Decay during the ageing of breadcrumb where both water migration and starch retrogradation occur concurrently is shown in Figure 2 for the "outer crumb". The FIDs were normalised by the signal magnitude measured using the solid echo pulse sequence in order to account for the impact of signal decay during the instrument dead time (~9μs).

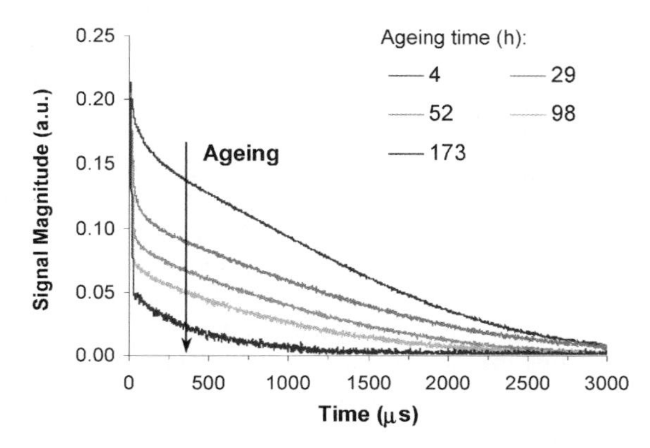

Figure 2 *Normalised FIDs of outer crumb samples after different ageing times.*

The results clearly show 3 main changes:

- Decrease of the total signal magnitude reflecting the loss of water from the crumb to the crust and the surrounding atmosphere (see Figure 1).
- Decrease of the liquid-like component of the FID reflecting the loss of water mentioned above and the reduction in mobility of a fraction of the bread which becomes part of the rigid component of the FID (this was discussed by Farhat et al.[6,11,13]).
- Decrease in the relaxation time of the liquid-like component resulting from the combined impact of reduced mobility of the bread matrix and the loss of the main mobile component, water.

A rough quantification of the changes shown in Figure 2 was obtained by plotting the changes in the signal magnitude at 100μs, i.e. beyond the relaxation of the rigid component of the FID. The results for the 3 parts of the baguette samples are shown in Figure 3, which, to a large extent, mirrors the measured changes in water content (Figure 1). While the results are expected from a direct water content – mobile FID component equivalence, they should be considered with interest as they suggest that the main phenomena controlling the changes in TD-NMR data during bread staling studies are changes in overall and local water contents. This is somehow in contradiction with many studies performed at constant water contents [6,10-12] and explains others, which overlooked the impact of water content changes in their interpretations.

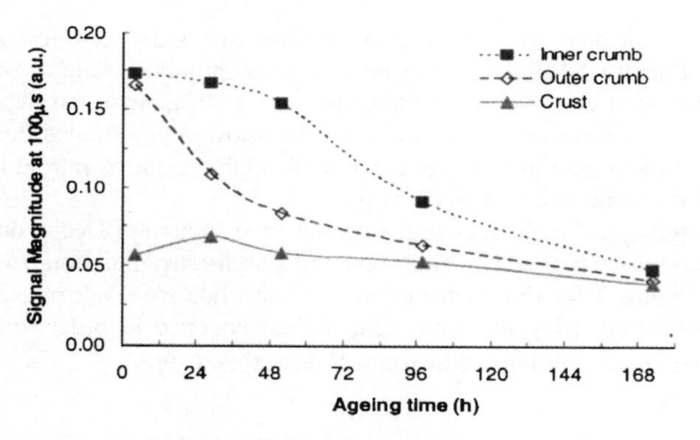

Figure 3 *FID signal magnitude at 100μs for the different parts of baguette samples as a function of ageing time (signal magnitudes were normalised by the corresponding initial solid echo signal).*

CPMG decays were acquired and analysed using a distributed exponential algorithm (using Oxford Instruments WinDXP software) for a further insight into the behaviour of water in the bread and its changes during ageing. It was not possible to acquire CPMG decays on the crust as most of the FID signal had decayed by the time the 180° was applied (τ=200μs) due to the low water content. Figure 4 shows the spin-spin relaxation spectra for the inner and outer crumb fractions. The spectra were analysed for $T_2 \geq 400\mu s$ i.e. no T_2 values shorter than the first measured point on the CPMG (2τ=400μs) was attempted. At short ageing times, a proton population centred at around ≤400μs was discerned for both the inner and outer crumb fractions. Except for this minor population, the relaxation spectra exhibited one main peak which at short storage times (4h) was centred at ~9.3ms for the inner crumb and 7.1ms for the outer crumb. This proton population reflects essentially the dynamic of water. The relatively short T_2 values reflect the fast proton exchange between the hydroxyls of water and those of the starch and the restricted water mobility within the polymers matrix (starch and glutens). The T_2s reduced to ~1ms after 173h of storage for both crumb fractions.

The CPMG relaxation spectra were normalised relative to the magnitude of the FID measured using the solid-echo pulse sequence and thus one can compare profiles acquired at different ageing times and also can directly compare the results of the inner and outer crumb.

In line with the water content results (Figure 1), the outer crumb showed both a smaller CPMG signal and a lower T_2 reflecting the lower water content and the subsequent lower water mobility usually observed in hydrophilic systems as water content decreases. Furthermore, the relaxation spectra were in agreement with the water content results in that the changes in spin-spin relaxation of the inner crumb were more gradual than those in the outer crumb.

These findings confirmed the observations discussed in Figures 2 and 3 and raised the same questions regarding whether starch retrogradation, which would be expected to occur during bread staling, played any significant role in controlling the NMR relaxation properties in bread systems as was demonstrated for starch systems [6,11-13]. This is addressed in the next section of this chapter.

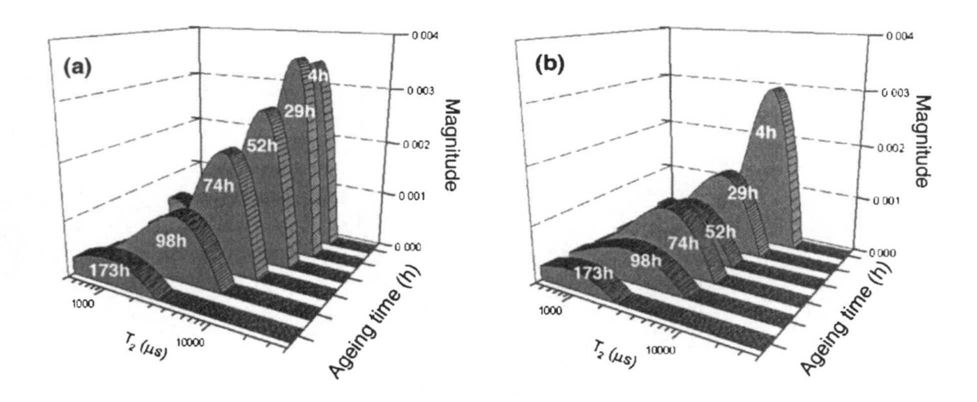

Figure 4 *Water relaxation spectra in the inner (a) and outer (b) crumbs as a function of ageing time.*

3.3 The relative impacts of the reduction of water content and starch retrogradation to the decrease in molecular mobility recorded during bread staling

In addition to the main experiment where water content and extent of retrogradation of the crumb changed during ageing, the following carefully controlled experiments were performed:

(i) Baguette outer crumb was stored at constant water content (~41% water) at 20°C and changes relating to retrogradation were monitored over 7 days (Figure 5b).

(ii) Outer crumb retrograded for 98h at constant water content (see above) to achieve "maximum" retrogradation as verified by Differential Scanning Calorimeter and

Wide Angle X-Ray Diffraction (results not shown), was equilibrated to a range of water contents by storing over different RHs (Figure 5c).

The CPMG decays were analysed to a distribution of exponentials as was described above. The T_2 values corresponding to the main relaxation population are shown in Figure 5 (similar results were obtained from the FID magnitude data).
The results indicated that although retrogradation (at constant water content) does contribute to a decrease of water mobility as is evident in the insert of Figure 5, this decrease of around 1ms over 173h of storage, is small compared to the change measured when both the water content and the extent of retrogradation varied simultaneously. It is not surprising that even in the case of a "maximally retrograded" sample (Figure 5c), water content had a significant effect on the measured T_2.

The results of the inner and outer crumb where both water content and extent of retrogradation varied where included to demonstrate that water content was the dominant parameter thus the results from the 2 fractions of the baguettes fall on a "master-curve" when plotted as a function of water content, even if the points acquired at specific ageing times do not overlap (different water contents).

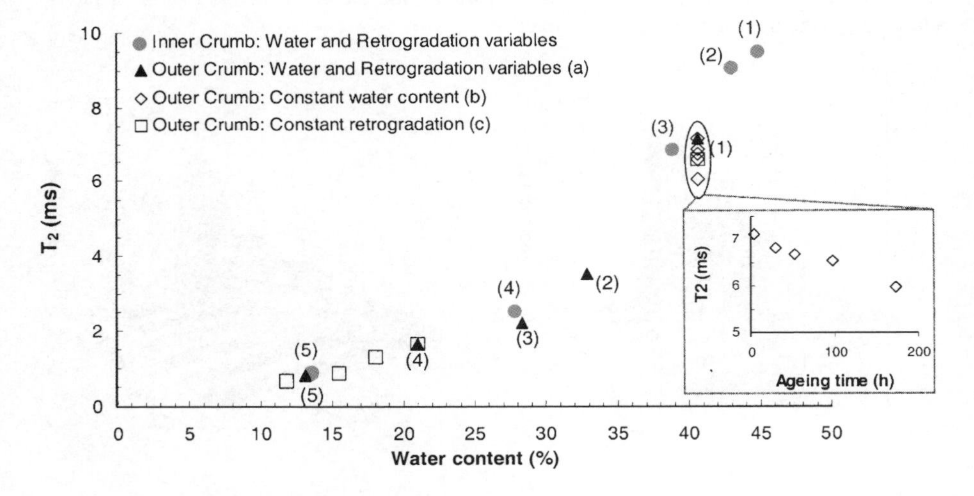

Figure 5 *Relaxation map: T_2(CPMG) in the baguette crumb during ageing at 54% RH. The ageing times for the inner (•) and outer (▲) crumbs were 4h (1), 29h (2), 52h (3), 98h (4) and 173h (5).*
The insert shows the changes observed during retrogradation of the outer crumb at constant water content (b)

4 CONCLUSIONS

The processes of water migration (throughout the crumb, from the crumb to the crust and loss to the surrounding) and starch retrogradation were systematically characterised in commercial baguette style breads.

The reduction in molecular mobility of both the bread matrix (probed through the characteristics of the FID decay) and of water (probed through the characteristics of the CPMG decay) are dominated by the changes in overall bread water content and within its different regions.

These results have several implications:

- They raise the question regarding the reasons behind the difference in the impact of retrogradation at constant water content on molecular dynamics between pure starch systems and bread[13]. Retrogradation leads to much more pronounced changes in NMR relaxation parameters in starch systems than in bread[6,13].
- They highlight the risks of artefacts when studying the ageing of complex foods where composition (in this case water content) varies across the food and with time and can impact the kinetics of other processes such retrogradation.
- The importance of not overlooking the role played by water when designing any antistaling solution.

References

This is only a very small subset of the considerable volume of literature on the applications of nuclear magnetic resonance techniques to study bread staling and related phenomena.

1. R.C. Hoseney, in *Physical Chemistry of Foods*, ed. H.G. Schwartzberg and R.W. Hartel, Marcel Dekker, Inc., New York, 1992, p. 443.
2. A. Schiraldi, L. Piazza and M. Riva, *Cereal Chem.*, 1996. **73**, 32.
3. L. Manzocco, M.C. Nicoli and T.P. Labuza, *Ital. J. Food Sci.*, 2002. **14**, 235.
4. Y. Vodovotz, E. Vittadini and J.R. Sachleben, *Carbohydr. Res.*, 2002. **337**, 147.
5. M.A. Ottenhof and I.A. Farhat, *Biotechn. and Gen. Eng. Rev.*, 2004. 21: p. 215.
6. I.A. Farhat, J.M.V. Blanshard, and J.R. Mitchell, *Biopolymers*, 2000. 53, 411
7. J.A. Gray and J.N. Bemiller, *Comp. Rev. in Food Sci. and Food Safety*, 2003. 2, 1.
8. A.C. Eliasson, in *New approaches to research on cereal carbohydrates*, ed. R.D. Hill and L. Munck, Elsevier, Amsterdam, 1985, p. 93.
9. J.A. Troller and J.H.B. Christian, in *Water activity and food*, Academic Press, New York, 1978, p. 13.
10. A. Schinardi and D. Fessas, in *Bread Staling*, ed. P. Chinachoti and Y. Vodovotz, CRC Press, 2001, p. 1.
11. I.A. Farhat and J.M.V. Blanshard, in *Bread staling*, ed. P. Chinachoti and Y. Vodovotz, CRC Press, 2001, p. 163.
12. I.A. Farhat, In: *The Stability and Shelf-life of Food, Eds Kilcast D and Subramaniam P.J, CRC Press* 2000, p. 129
13. I. A. Farhat, M.A. Ottenhof, V. Marie and E. de Bezenac, In: *Magnetic Resonance in Food Science: Latest Developments*, 2003; p 172

HIGH RESOLUTION NMR TOOLS FOR THE ANALYSIS OF BEER AND WINE

A. M. Gil and J. Rodrigues

Department of Chemistry and CICECO, University of Aveiro, Campus Universitário de Santiago, 3810-193 Aveiro, Portugal

1 INTRODUCTION

Nuclear Magnetic Resonance (NMR) spectroscopy has long been recognised as a promising tool for the rapid compositional profiling of liquid foods. High resolution 1D NMR spectroscopy has the valuable ability to register the spectral profiles characteristic of compounds present in a mixture in concentrations higher that a few milligrams per litre. This results in a complex overall NMR profile, characterised by extensive signal overlap and posing the problem of how to extract valid compositional and structural information from such complex profiles. Provided that the complex 1D NMR spectra of mixtures such as beer, wine, juices may be unravelled and converted into a list of compounds and concentrations, this information is extremely valuable as a fingerprint of the food. Such information may then be used to monitor deviations relatively to a profile defined as acceptable or normal, in a number of different practical situations.

Interpretation of the NMR spectra of complex mixtures may be carried out using different tools, depending on the specific aim of the study and on the type and number of samples available. Generally, in the last few years, developments in the use of NMR spectroscopy for liquid food analysis have proceeded along two main lines: 1) the application of new methods to allow increasingly detailed and sensitive qualitative analysis and 2) the application of NMR in tandem with multivariate analysis for quantitative and semi-quantitative analysis, increasingly tackling specific practical industrial problems.

This chapter will start by presenting an overall view of the state-of-the-art NMR methods for qualitative analysis of liquid foodstuffs. These include simple variations of the standard 1D and 2D NMR experiments, in order to broaden the range of compounds detected, hyphenated NMR (LC-NMR and LC-NMR/MS) and diffusion-ordered spectroscopy (DOSY). The latter methods will be discussed regarding their potential as well as their difficulties/disadvantages and the examples given will refer to applications to beer and wine.

The second part of this text will address the applicability of high resolution NMR to specific problems in the food industry. This will necessarily entail the analysis of large numbers of samples and validation of the results making use of multivariate analysis. This strategy allows for composition fluctuations to be detected and correlated to factors such as origin and history of sample. The examples given relate to the monitoring of beer quality at different production sites and at different dates and for the monitoring of wine composition as a function of harvest year and corresponding climatic factors.

2 METHODS

Beer bottles were opened immediately before NMR analysis. All samples were subjected to ultrasonic degassing (10 minutes). For NMR spectroscopy, beer samples were prepared as to contain 10% (v/v) D_2O and 0.02% (v/v) 3-(trimethylsilyl)propionate sodium salt (TSP, used as chemical shift and intensity reference). For hyphenated NMR analysis, samples were simply degassed. Details of the NMR experiments may be found elsewhere for high resolution NMR[1-3], DOSY[4] and hyphenated NMR[5,6]. Details of the multivariate analysis methods applied to the beer NMR data may be found in references 1 to 3.

Wine bottles were also opened immediately before NMR analysis. For NMR spectroscopy, all samples were prepared as to contain 10% (v/v) D_2O and 0.002-0.05% (w/v) TSP. Details of standard NMR and DOSY experiments may be found in reference 7.

3 RESULTS AND DISCUSSION

3.1 Qualitative analysis

Beer and wine are both good examples of complex mixtures which comprise many different families of compounds in a wide range of concentrations. Their potential as sources of interesting natural compounds, e.g. with antioxidant or antimicrobial properties, is clearly recognised and, together with the quest for compounds which may act as quality indicators, this justifies the efforts in lowering detection limit and/or the degree of spectral overlap. This should help in the build-up of more complete NMR-based composition tables which may be established as characteristic of each food.

Figure 1a shows a typical standard 1D NMR spectrum (500 MHz) of a beer. The assignment of most peaks relies necessarily on the comprehensive analysis of 2D NMR homonuclear and heteronuclear experiments, such as total correlation spectroscopy (TOCSY), $^1H/^{13}C$ correlation spectroscopy and *J*-resolved 2D spectroscopy. Detailed interpretation of these experiments is discussed elsewhere[1-2] and an up-to-date table of assignments for beer may be found in reference 2. For wine, a similar strategy has been applied, enabling the identification of many tens of compounds[8,9]. Going back to the case of beer and the spectrum in Figure 1a, it is noted that, in the higher field region, from 0 to 3 ppm, signals arising from aliphatic compounds or moieties and comprising aliphatic amino acids, organic acids and alcohols are observed. In the mid-field region, from 3 to 6 ppm, the signals arise mainly from carbohydrates and some contributions from organic acids. Of these, dextrins (linear or ramified oligomers of glucose) are the most abundant compounds, typically present in concentrations between 2 and 3 g/100ml. Finally, the spectral region from 6 to 9 ppm is the lowest intensity region and shows peaks arising from aromatic amino acids, organic acids and nucleotide bases. In spite of the use of several 2D NMR experiments, at least 20 compounds still remain unassigned in beer due to incomplete spin system identification (mainly due to signal overlap) or to inherently spin system overlap due structural similarity of the compounds. Dextrins are an example of the latter situation and only average values for oligomer size and branching degree may be obtained from the 1D NMR spectrum. Other examples of incomplete or ambiguous assignments relate to a) adenosine and inosine, b) phenylalanine and 2-phenylethanol and c) tyrosine and tyrosol. In beer, the distinction between the aromatic amino acids and corresponding alcohols is particularly important, since the degree of amino acid conversion into their alcoholic forms reflects the extent of the fermentation process[10].

As a first additional approach to improve signal assignment, simple spectral editing tools may be used in order to separate signals of compounds with lower M_r (using a Carr Purcell Meiboom Gill pulse sequence) or higher M_r (using a diffusion filter, with basis on the lower diffusivity of larger molecules). Figures 1b and c show, respectively, the result of those experiments when applied to beer. In Figure 1b, the broad underlying components are lost thus clarifying some spectral profiles, particularly in the high- and low-field regions. On the other hand, Figure 1c shows that in the diffusion-edited spectrum practically all narrow signals are lost, leaving broad profiles corresponding to lipids, dextrins and, possibly, polyaromatics. Although these results provide relatively little advance in specific assignment of new signals, they may be further pursued, for instance, by coupling the spectral editing pulse sequences with homonuclear 2D experiments, in order to enable the assignment of particular spectral profiles made clearer in the edited experiments. This has been carried out successfully for biofluids[11], mixtures of at least as much complexity as foods.

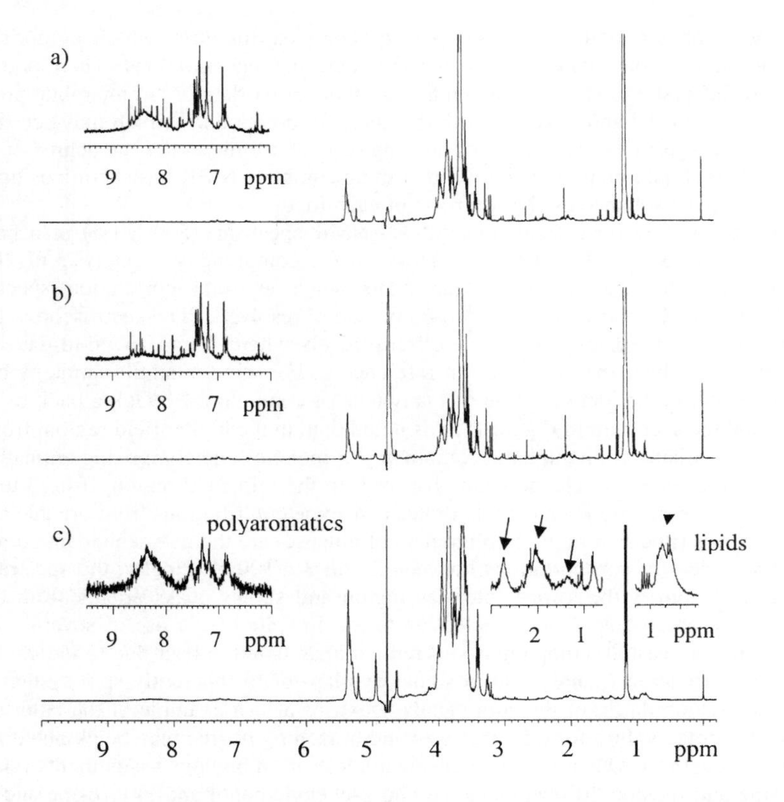

Figure 1 *Typical ^1H NMR spectra (500 MHz) of a beer sample using a) a standard pulse sequence, b) a CPMG-based sequence to select the signals from lower M_r compounds and c) a diffusion filter to select the signals from higher M_r compounds.*

The difficulties arising from signal overlap may be significantly resolved using hyphenated NMR methods. LC-NMR/MS has been applied to the characterisation of aromatic compounds and carbohydrates in beer[5,6]. In the former case, the main achievement was the clear distinction of the aromatic amino acids and the corresponding alcohols, which elute at very different times and show clear non-overlap spectral profiles. The quantification of each of the fractions would be of importance in the evaluation of fermentation extension in the brewing process and is a possible direction for future work. In the case of beer carbohydrates, several different dextrins could be identified and their size and branching degree determined. Figure 2 shows how a simple and rapid (*ca.* 30 minutes) LC-NMR run (using continuous flow) may provide qualitative information on the carbohydrate composition of beers. The nature of beer dextrins (size and branching degree) is a good indicator of the conditions of the malting and mashing steps of beer production and, thus, a qualitative LC-NMR screening showing differences such as those in Figure 2 may be of value in malting/mashing control. However, the use of hyphenated NMR methods does involve significant costs both in equipment and in deuterated solvents and, if sensitivity improvement is required, more lengthy experiments are required compared to the rapid simple LC-NMR run shown in Figure 2. At this stage, the use of these methods is therefore limited to a few selected samples of interest and not applicable if high sample throughput is required.

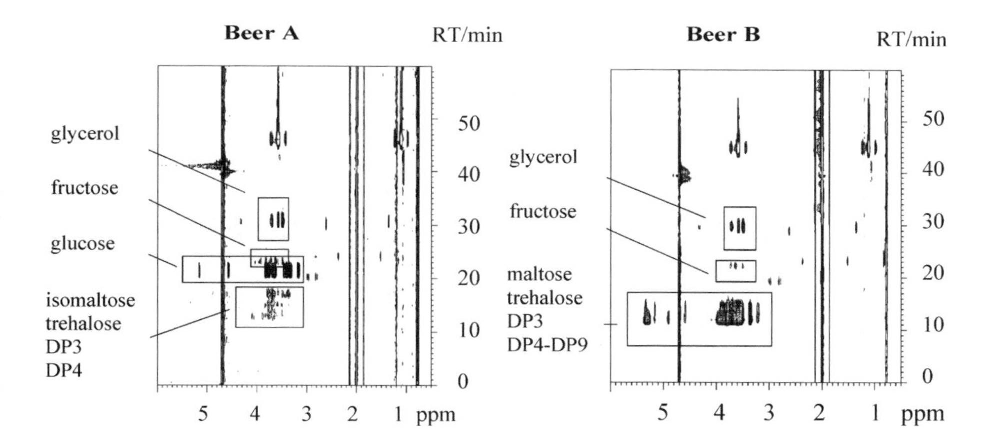

Figure 2 *LC-NMR continuous flow plots obtained for two different beers. Beers A and B are, respectively, richer in lower and higher M_r glucose oligosaccharides.*

Diffusion ordered spectroscopy (DOSY) is a more economic valuable tool to disentangle the NMR signals in a 1D spectrum and achieve their assignment. The theoretical principles of this method are described in detail elsewhere[12] and the experiment results in the separation of subspectra as a function of the diffusivity of the corresponding compounds. Plots such as those shown in Figure 3 are obtained, each row showing a subspectrum corresponding to a certain diffusivity value. The examples shown relate to the aromatic composition of two Port wine samples differing in age. The results clearly show that the large aromatics present in the young wine are not present in the older wine, reflecting condensation and precipitation of anthocyanin-based polymers[7], as expected with basis on wine chemistry. Furthermore, several new compounds are detected in the older wine, indicated as peaks 1 to 5 in Figure 3a, compounds 3 and 4 corresponding to singlets and initially remaining unassigned. Identification of unknown compounds may be attempted using known compounds as calibrants for diffusivity. For instance, formic acid (signal 1), ethanal (signal 2) or *p*-hydroxybenzoic acid (signal 5) have clear and well resolved DOSY peaks for which diffusivity values may be read with relative precision. A simple calculation based on the scaling of the diffusion coefficient with the cube root of relative molecular mass, M_r, will give the predicted molecular weight of the unknown compounds and an assignment may be put forward on this basis. In this way, it could be concluded that signals 3 and 4 in Figure 3a correspond to compounds with 190 and 70 molecular weight, respectively, the latter possibly corresponding to ethyl formate. This shows that detailed information may be obtained about chemical changes taking place in foods, using the DOSY method. This method enables the assignment of singlets (of particularly difficult interpretation by NMR) and may potentially also detect complexation or bonding between different moieties e.g. carbohydrate and aromatic. However, precision in the diffusivity dimension may only be achieved for NMR signals free of overlap and this may be a stumbling block in its direct application to complex mixtures.

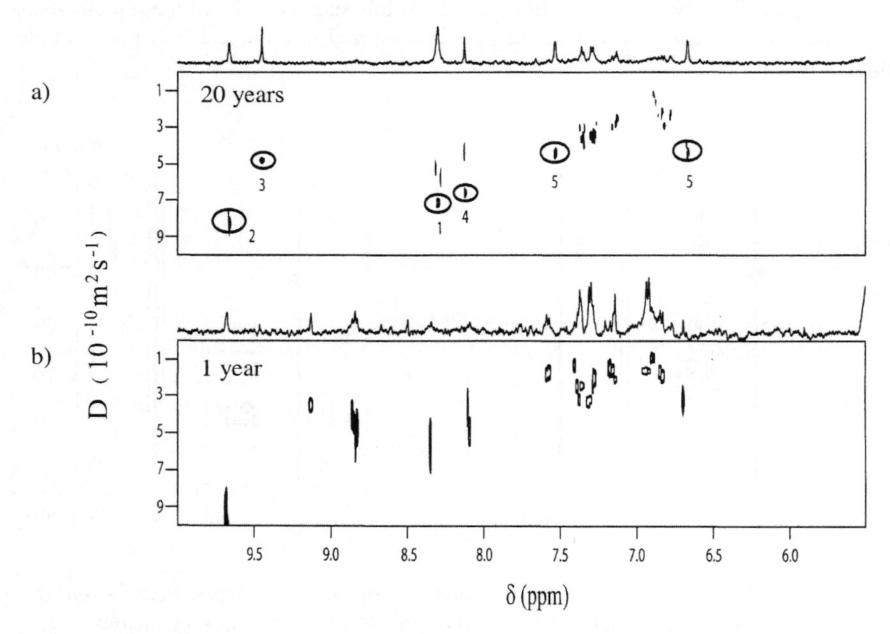

Figure 3 *Aromatic regions of the DOSY plots obtained for two Port wine samples differing in age a) 20 years and b) 1 year (figure adapted from reference 7).*

The tandem use of hyphenated NMR and DOSY is, probably, the most likely direction to follow if further qualitative insight is to be gained on liquid foods by NMR. Whereas standard 1D and 2D NMR provide information on few mg/L concentrations, hyphenated NMR improves sensitivity by about 30%. Further lowering of the detection limit should require the use of cryoprobe facilities and concentrations down to μg/L should then be accessed.

3.2 Quantitative and semi-quantitative analysis – the use of multivariate analysis

Once the qualitative analysis is deepened to the appropriate level for each specific problem, it will most often be of interest to establish quantitative or semi-quantitative conditions in order to apply the NMR-based tool to routine and high throughput sample analysis. This requires necessarily an analytical strategy that can handle large numbers of samples so that any variation in concentration may be evaluated and identified as acceptable or not acceptable.

Quantitation by NMR spectroscopy traditionally relies on the evaluation of signal areas, usually relatively to that of a signal arising from an internal standard compound (e.g. TSP). In complex mixtures, however, this procedure is prone to large errors arising from changes in peak areas due to interaction between the standard compound and component(s) in the sample. This method has been performed for a large set of beers and compared to an alternative method based on the multivariate method Partial Least Squares (PLS)[3]. The latter method involved the parallel compound quantification (for organic acids and amino acids) by a reference method, e.g. capillary electrophoresis or liquid chromatography, in order to calibrate the NMR data. The PLS-based approach was shown to be more suitable to handle large number of samples and enabled more precise quantitation, compared to the traditional direct integration method. In order to explore the applicability of the NMR/PLS quantitation approach to routine analysis, the robustness of the model must necessarily be validated, not only for instrumental fluctuations in the same NMR instrument, but also for use in different spectrometers (at both same and different field strengths).

The multivariate analysis method of Principal Component Analysis (PCA) may also be useful in semi-quantitative screening of samples in order to look for deviations in concentrations of selected compounds taken as indicators for certain steps of the production process. In the case of beer, this has been applied to determination of beer type and origin[1,2]. The most recent study has addressed a set of beers of the same label and type, produced in different countries and at different dates, with the aim of establishing beer reproducibility between sites and dates. The method found most suitable was the separate PCA of spectral subregions (0-3 ppm, 3-6 ppm and 6-9 ppm), after a degree of spectral linebroadening (5-10 Hz) or bucketing. This pre-analysis processing was found necessary to minimize peak shifts due to slight pH differences in the samples which would, otherwise, increase sample dispersion in the PCA plot considerably. Figure 4 shows the PCA scores plot obtained for the higher field regions of all spectra (0-3 ppm).

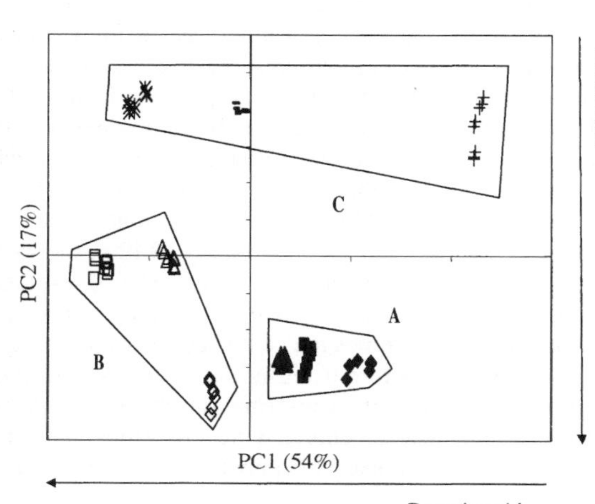

Figure 4 *PCA scores plot obtained for the aliphatic regions of the NMR spectra of beers produced in different countries (A,B,C) and at different dates (♦A1, ■A2, ▲A3, ◊B1, □B2, ΔB3, xC1, –C2, +C3).*

Considering the corresponding loadings plot (not shown), it was found that PC1 and PC2 were respectively related mainly to the concentrations of pyruvic acid and lactic acid. It is clear that beers can be grouped in terms of their country of origin and that reproducibility in time is best in site A and poorer in site C. Higher contents in both pyruvic and lactic acids (as is the case of site B beers) may be indicative of poorer yeast quality or older yeast generation[2], this property being most efficiently controlled in site A.

The same simple analytical approach was applied to the mid-field and lower-field regions of the spectra thus giving information about variations in dextrin content/nature and in aromatic compounds. Interestingly, considerable separation of beers may always be achieved in terms of their production sites, showing differences in dextrin branching degree and in the contents of compounds such as tyrosine and/or tyrosol, uridine and adenosine and/or inosine. Such differences may be interpreted as reflecting the temperature conditions during malting and mashing of beer and the extension of the fermentation process[2]. Such characteristics may, in this way, be rapidly monitored through the NMR/PCA approach and this may be of potential use in routine beer analysis and process monitoring.

A finer challenge is the study of the brewing process reproducibility between sites and dates within one same country, since in such case all factors like raw materials, water, plant origin and process recipe are much more closely controllable. However, initial unpublished results have shown that significant fluctuations still occur in the concentrations of specific compounds which may be correlatable to the exact and fine characteristics of equipment and procedure which differ slightly between sites. The extent of these effects is being evaluated so that the applicability of the procedure in the factory may be considered.

A second and final example of the NMR/multivariate analysis procedure relates to changes in wine composition due to geographical and climatic effects over the years[13]. This study follows similar previous work for the classification of wine according to vine variety, vintage or geographical origin[14,15]. Here, a set of Cabernet Sauvignon wines produced in southern Brazil was addressed and the study entailed the analysis of a large number of wine samples to evaluate not only the changes between harvest years, but also the range of compositional changes for one same harvest. Preliminary NMR/PCA results have shown significant changes in alcohols, acetates, acetoin, some organic acids and aromatic compounds, as a function of harvest year and for the same production site. The concentrations of organic acids and aromatic compounds seem to be more sensitive to geographical effects, as seen for wines produced in different regions. A more thorough understanding of the origins of these changes may be obtained through the application of PLS in order to evaluate any correlations between NMR data and the climatic factors registered over the period that includes the several harvests under study. Within the long list of climatic factors, the most significant correlations were found for a) average temperature, b) average and total sunlight and c) average relative humidity. This last correlation seems particularly interesting since it overcomes the contributions from average or total rainfall, initially believed to be important factors determining wine composition. These preliminary results suggest that additional important information about the role of climatic properties on wine composition (and, hence, general wine properties) may be obtained through NMR and multivariate analysis methods. This may become the basis of a useful predictive strategy in wine production.

4 CONCLUSIONS

Qualitative direct 1D/2D NMR analysis of liquid foods, at a field strength of 500 MHz, achieves detection of compounds down to few mg/L in concentrations. Many tens of compounds are detectable in this way, but many other compounds remain unidentified due to difficulties arising from the strong signal overlap in the 1D and 2D spectra. Hyphenated NMR, particularly LC-NMR/MS, has proved useful in unravelling further compounds in the carbohydrate and aromatic regions of the spectra. The costs and length of some of the experiments must, however, be considered when practical applications are to be explored. In addition, the DOSY experiment adds diffusivity information to NMR profiles thus aiding considerably in the assignment of unknown compounds. This is, however, only achieved successfully in the case of non-overlapped NMR profiles, which is not always the case in complex mixtures. A tandem use of LC-NMR and DOSY is recommended as a most efficient strategy in the assignment of compounds in liquid foods.

Quantitation and semi-quantitative analysis of large sets of samples are necessary in order to explore the uses of NMR-based methods in the food industry. Quantitation of selected compounds has been found best achievable using PLS methods, however, the robustness of the resulting NMR/PLS models is still to be evaluated for distinct instrumental and sample natural fluctuations. Semi-quantitative models such as NMR/PCA models are of significant value in the rapid detection of concentration deviations from an established normal range. This approach has been tested for beers produced in different countries and for beers produced in different sites of one same country. The fluctuations detected in the concentration of specific compounds are related to particular conditions in different steps of the beer production process and may be used to monitor their performance. The same strategy seems promising in the monitoring of wine characteristics and production climatic conditions. These examples show that the relatively simple tandem use of NMR and multivariate analysis may be of practical use in the food industry, as long as the difficulties arising from the handling of large sets of samples are resolved. In addition to simple methods such as pre-processing with linebroadening or bucketing routines, the expertise and specificity of some of the newly developed multivariate methods for biofluids[16-18] should be of great value in the food analysis area.

Acknowledgements

The authors would like to acknowledge I.Duarte, C.Almeida and A.Barros from the University of Aveiro, Portugal, G.Morris and M.Nilsson from the University of Manchester, M.Maraschin from the University of Santa Catarina, Brasil, and Manfred Spraul from Bruker BioSpin GmbH, Germany.

Funding is gratefully acknowledged from the Foundation for Science and Technology, Portugal; Bruker Biospin, Germany; the Royal Society of Chemistry, London; CAPES, Brazil; Unicer-Bebidas de Portugal.

References

1 I. F. Duarte, A. Barros, C. Almeida, M. Spraul and A. M. Gil, *Journal of Agricultural and Food Chemistry*, 2004, **52**, 1031.

2 C. Almeida, I. F. Duarte, A. Barros, J. Rodrigues, M. Spraul and A. M. Gil, *Journal of Agriculture and Food Chemistry*, 2006, **54**, 700.

3 L. Nord, P. Vaag and J. Duus, *Analytical Chemistry*, 2004, **76**, 4790.

4 A. M. Gil, I. F. Duarte, E. Cabrita, B. Goodfellow, M. Spraul and R. Kerssebaum, *Analytica Chimica Acta*, 2004, **506**, 215.

5 A. M. Gil, I. F. Duarte, M. Godejohann, U. Braumann, M. Maraschin and M. Spraul, *Analytica Chimica Acta,* 2003, **488**, 35.

6 I. F. Duarte, M. Godejohann, U. Braumann, M. Spraul and A. M. Gil, *Journal of Agricultural and Food Chemistry*, 2003, **51**, 4847.

7 M. Nilsson, I. F. Duarte, C. Almeida, I. Delgadillo, B. Goodfellow, A. M. Gil and G. A. Morris, *Journal of Agricultural and Food Chemistry*, 2004, **52,** 3736.

8 I.J. Kosir and J. Kidric, *Journal of Agricultural and Food Chemistry*, 2001, **49**, 50.

9 I.J. Kosir and J. Kidric, *Analytica Chimica Acta*, 2002, **458**, 77.

10 P. S. Hughes and E. D. Baxter, *Beer: Quality, Safety and Nutritional Aspects*, Royal Society of Chemistry, Cambridge 2001.

11 H. R. Tang, Y. L. Wang, J. K. Nicholson and J. C. Lindon, *Analytical Biochemistry*, 2004, **325**, 260.

12 G. A. Morris, *Encyclopaedia of Nuclear Magnetic Resonance,* John Wiley & Sons, Chichester, 2002, Volume 9, p. 35.

13 M. Maraschin, A. Barros and A. Gil, unpublished results.

14 J.T.W.E. Vogels, A. C. Tas, F. van den Berg and J. ven der Greef, *Chemometrics and Intelligent Laboratory Systems*, 1993, **21**, 249.

15 M.A. Brescia, I. J. Kosir, V. Caldarola, J. Kidric and A. Sacco, *Journal of Agricultural and Food Chemistry*, 2003, **51**, 21.

16 J. Forshed, I. Schuppe-Koisyinen and S. P. Jacobsen, *Analytica Chimica Acta*, 2003, **487**, 189.

17 R. Stoyanova, A. W. Nicholls, J. K. Nicholsom, J. C. Lindon and T.R. Brown, *Journal of Magnetic Resonance*, 2004, **170**, 329.

18 B. M. Beckwith-Hall, J. T. Brindle, R. H. Barton, M. Coen , E. Holmes, J. K. Nicholson and H. Antti, *Analyst*, 2002, **127**, 1283.

ADULTERATION STUDY IN BRAZILIAN HONEY BY SNIF AND ^{1}H NMR

E.F. Boffo[1], L.A. Tavares[1], A.G. Ferreira[1], M.M.C. Ferreira[2] and A.C.T. Tobias[3]

[1] Departamento de Química, Universidade Federal de São Carlos, São Carlos - SP, Brasil.
[2] Instituto de Química, Universidade Estadual de Campinas, Campinas - SP, Brasil.
[3] Centro Regional Universitário Espírito Santo do Pinhal, Espírito Santo do Pinhal - SP, Brasil.

1 INTRODUCTION

The apiculture in Brazil are increasing in the last years and will be possible to ingress in the set of the honey exportation country selling to Europe, United States and Japan[1]. However, it is very important to have a good quality control to attend the exigency of the consuming market. Therefore, a more rigorous quality control must be carried to increase its competitively.

The food products authentication is a primary importance and from commercial rules point of view, quality standards have been established through the requirement of quality labels that specify the product chemical composition, validate/expire data, origin country and in some case the specific place that they are produced and/or collected in a particular country. From the economic point of view, product authentication is essential to avoid unfair competition[2]. Besides this, different countries have they property legislation about the quality control which difficult the standardisation. For instead in Brazil the legislation, according to resolution no.11/2000, forbidden the use of any additive in the honey which is characterised like a natural, sweet substance produced by *Apis mellifera* bees from nectar of plants or from secretions of living parts of plants or excretions of plant-sucking insects on the living parts of plants, which the bees collect, transform by combining with specific substances of their own, deposit, dehydrate, store and leave in honeycombs to ripen and mature[3].

Site-specific Natural Isotope Fractionation Nuclear Magnetic Resonance (SNIF-NMR) is one of the best answer for the question of adulterations when are involved the product biosynthetic origin. Stemming from the work of Gérard J. Martin and colleagues at Nantes University, France, SNIF-NMR is probably the most applied and specific method for food product authenticity. After the technique was developed in the early 1980s, the original application of it was the detection of the chaptalization or enrichment, in wines[4]. Since then, SNIF-NMR has been extended to other products than wine to show the addition of undeclared sugars[4-7]. Martin et al. applied SNIF-NMR alone[8] or SNIF-NMR and stable isotope ratio analysis mass spectrometry (SIRA-MS)[9] to detect added sugar and to evaluate authentication in fruit juices.

SNIF-NMR techniques, which determine the ^{2}H/^{1}H relationship, provides an isotopic criteria to characterise a biochemical transformation such as fermentation and it is enable to measure the isotopic ratios for the end products which can be correlated with the precursors[10]. Therefore, it is possible to verified the type of plant where the nectar was collected by the bees, if they use the cycle of carbon fixation in their biosynthetic mechanisms C_3 and C_4[11, 12] which result in a different isotopic ratios. It also allows to inquire if the honey was adulterated with some another type of sugar.

With this objective, ethanol produced from fermentation of honey from different source were analysed. The ethanol molecule possess three mono-deuterated isotopomers at natural abundance[13] $CH_2\mathbf{D}CH_2OH$ (I), $CH_3CH\mathbf{D}OH$ (II) and $CH_3CH_2O\mathbf{D}$ (III), therefore, the deuterium can be located either on the methyl (I), the methylene (II) or the hydroxyl (III) sites. Because of the low natural abundance of deuterium relative to hydrogen, the probability to find a bi-deuterated species is very low and can be ignored[14].

Honey adulteration is easily carried out by several ways like a simple sucrose addition (of a C_4 plant like sugar cane), a mixture of honey from different source and more recent, to feeding the bees with a sucrose solution or sucrose syrup with vegetal extract when they are harvesting the nectar. The deuterium/hydrogen ratios measured at the methyl site of ethanol $(^2H/^1H)_I$ in honeys differ significantly according to the nectar origin. By feeding the bees with sucrose solutions from sugar-cane, for instance, will increase the $(^2H/^1H)_I$ value if compared with that produced from C_3 plant, like nectar from eucalyptus or citrus[12].

This work describe the use of SNIF-NMR techniques to classified honey commercial samples using the $^2H/^1H$ isotopic relation of the methyl and methylene groups in ethanol by NMR. In this context, we can confirm the honey precedence and we can investigate the product adulteration. ^1H NMR was also used together with chemometric methods, Hierarchical Cluster Analysis (HCA), to complement this study. The HCA method examines the distances between all of the samples and represents the information in the form of a two-dimensional plot called a dendrogram[15].

2 MATERIALS AND METHODS

2.1 Samples

Forty honey samples obtained from the flowers of different plants like: citrus (*Citrus sp.*), eucalyptus (*Eucaliptus sp.*), assa-peixe (*Vernonia sp.*), wildflowers and feeding the bees with a solution of sucrose (sugar-cane - *Saccharum sp.*) were studied. Some of these samples were provided by the beekeepers and the others were bought in the Brazilian shopping particularly in S. Paulo state. All the sample are collected in the years 2004 and 2005.

2.2 Fermentation

The sample (10 g) was introduced into a 500 mL glass flask at 24 °C. 5 g of baker's yeast, *Saccharomyces cerevisiae*, were added with a continuous stirring during 48 hours. At the end of fermentation the ethanol was distilled and the liquid collected with the boiling point in the range of 76-78 °C.

Samples for NMR measurements were prepared using 600 μL of distillate ethanol and 100 μL of tetramethylurea (TMU), 99.0 %, which was used as an internal pattern.

2.3 ^{2}H and ^{1}H NMR Spectra

All the ^{2}H and ^{1}H NMR spectra of ethanol were carried out on a Bruker DRX400 - 9.4 Tesla spectrometer without fluorine lock device and all measurements were made in triplicate in a 5 mm direct probe, maintaining the temperature constant at 298K and with the same data acquisition and processing.

The ^{2}H NMR data (61.4 MHz) were acquired using broadband proton decoupling, spectral width 983 Hz, memory size 10k, acquisition time 5.2 s, relaxation delay 2 s, rf pulse (90°) 17.5 μs and 1024 FIDs were accumulated. The FIDs were processed with zero-filling using an exponential multiplication associated with a line broadening 1.0 Hz and was made the automatic correction of the baseline.

The ^{1}H NMR data were acquired using spectral width 6410 Hz, memory size 64k, acquisition time 5.2 s, relaxation delay 2 s, rf pulse (90°) 10.5 μs and 16 FIDs were accumulated. The FIDs were processed with zero-filling using an exponential multiplication associated with a line broadening 0.3 Hz. and was made the automatic correction of the baseline. The internal reference for ^{1}H NMR spectra was TMS (tetramethylsilane).

2.4 Calculations

The isotopic ratios were determined from the methyl $(^2H/^1H)_I$ a..d methylene $(^2H/^1H)_{II}$ sites from ethanol molecule using deuterium/hydrogen NMR. The quantitative data were obtained by the automatic integration of the peaks from the sample and internal pattern (TMU).

The isotopic ratios at the two ethanol sites were determined according to the equation 1[16,17]

$$\left(\frac{{}^2H}{{}^1H}\right)_i^A = \frac{I_i^A}{I^P} * \frac{P^P}{P_i^A} * \frac{m^P}{m^A} * \frac{M^A}{M^P}\left(\frac{{}^2H}{{}^1H}\right)^P \quad (1)$$

where I_i^A and I^P are the areas of signal i of A and of the methyl signal of TMU in the ^{2}H NMR spectrum. P_i^A and P^P are the stoichiometric numbers of hydrogens at site i and in the TMU. M^A, m^A and M^P, m^P are the molecular weight and mass of the investigated compound, A, and the reference, respectively.

2.5 ^{1}H NMR Spectra for chemometric analysis

The samples for ^{1}H NMR were obtained dissolving 150 mg of honey in 450μL of D_2O. All the ^{1}H NMR spectra were obtained using a 5 mm probe with inverse detection, in triplicate and in the same parameter conditions of acquiring and processing data, using zgcppr pulse program, acquisition time 7.02 s, memory size 64K, spectral width 4664 Hz, relaxation delay 1.5 s, rf pulse 8.50 μs with 64 FIDs accumulated. In the processing were used 32k points and was made baseline automatic correction.

2.6 Chemometrics treatment

The software used was the Pirouette® version 2.02. The data matrix was built with 4666 variables (columns) and 120 spectra (lines – 40 samples in triplicate). The variable pre-processing was autoscaling, and samples transformations was normalisation and apply the first derivative. The method used was incremental linking with Euclidian distance.

3 RESULTS AND DISCUSSION

In figure 1, we can observe the ^{1}H and ^{2}H NMR spectra of ethanol obtained from fermentation of a wildflower honey, where can be visualised the signals of ethanol, TMU and water.

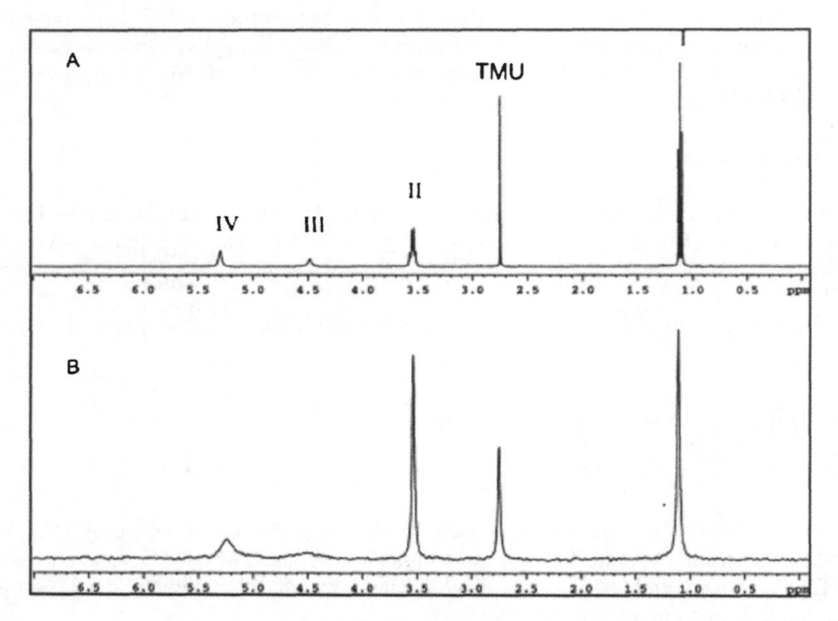

Figure 1 *Natural abundance NMR spectra of A) ^{1}H and B) ^{2}H of ethanol from a honey sample fermentation where I methyl, II methylene and IV hydroxyl from ethanol signals and III water*

The isotopic ratios $(^{2}H/^{1}H)_{I}$ and $(^{2}H/^{1}H)_{II}$ of ethanol samples obtained from honey fermentation are shown in table 1.

Table 1 *$^2H/^1H$ relations value for methyl(I) and methylene(II) sites of ethanol (in ppm)*

Honey	Isotopic Relation (ppm)**	
	$(^2H/^1H)_I$	$(^2H/^1H)_{II}$
Eucalyptus 1*	96.9 (± 0.1)	128.3 (± 1.1)
Eucalyptus 2	96.9 (± 1.0)	128.4 (± 0.2)
Eucalyptus 3*	97.3 (± 0.8)	129.6 (± 0.5)
Eucalyptus 4	97.9 (± 0.2)	128.2 (± 0.1)
Eucalyptus 5	99.2 (± 0.7)	127.5 (± 0.6)
Eucalyptus 6	99.4 (± 0.2)	128.8 (± 0.2)
Eucalyptus 7*	99.7 (± 0.8)	128.8 (± 0.7)
Eucalyptus 8	99.7 (± 1.0)	129.1 (± 0.5)
Eucalyptus 9	98.7 (± 0.5)	128.1 (± 0.5)
Eucalyptus 10	**112.2 (± 0.5)**	**128.0 (± 0.3)**
Citrus 1	100.3 (± 0.1)	127.6 (± 0.3)
Citrus 2	100.5 (± 0.3)	129.1 (± 0.5)
Citrus 3	100.7 (± 0.4)	128.0 (± 0.4)
Citrus 4*	100.7 (± 0.6)	128.2 (± 0.2)
Citrus 5*	100.9 (± 0.6)	127.3 (± 0.4)
Citrus 6	101.3 (± 0.9)	128.0 (± 0.5)
Citrus 7*	101.8 (± 0.3)	127.9 (± 1.1)
Citrus 8	101.8 (± 0.4)	128.1 (± 0.6)
Citrus 9	102.1 (± 0.2)	127.2 (± 0.2)
Citrus 10*	103.4 (± 0.2)	128.1 (± 0.9)
Citrus 11	**112.0 (± 0.3)**	**129.2 (± 0.5)**
Citrus 12*	102.4 (± 0.1)	127.6 (± 0.5)
Wildflower 1*	97.1 (± 0.4)	128.9 (± 1.0)
Wildflower 2	97.4 (± 0.4)	127.5 (± 1.0)
Wildflower 3*	98.0 (± 0.4)	128.0 (± 2.0)
Wildflower 4	98.5 (± 1.4)	127.4 (± 2.0)
Wildflower 5	99.4 (± 0.4)	127.7 (± 0.4)
Wildflower 6	99.4 (± 0.7)	128.5 (± 0.3)
Wildflower 7	99.8 (± 0.5)	128.2 (± 0.6)
Wildflower 8	100.1 (± 0.1)	127.5 (± 0.6)
Wildflower 9	100.4 (± 0.6)	127.6 (± 0.6)
Wildflower 10*	100.4 (± 0.7)	129.1 (± 0.7)
Wildflower 11	100.5 (± 0.4)	128.4 (± 0.4)
Wildflower 12	101.1 (± 0.6)	127.7 (± 0.1)
Wildflower 13	102.0 (± 0.1)	128.4 (± 0.6)
Wildflower 14	102.5 (± 0.4)	127.7 (± 0.4)
Wildflower 15*	103.5 (± 0.2)	129.3 (± 0.2)
Assa-peixe 1*	101.2 (± 0.1)	127.8 (± 0.3)
Assa-peixe 2*	101.7 (± 0.1)	127.9 (± 0.4)
Sugar-cane*	**111.8 (± 0.2)**	**128.0 (± 0.5)**

* standard samples.
** triplicate medium values

Analysing the isotopic relation values presented in the table 1 we can see that the value for methyl $(^2H/^1H)_I$ and for methylene $(^2H/^1H)_{II}$ from ethanol were kept in the range of 98-102 ppm and 127-129 respectively.

The isotopic relation values obtained for honey which were feeding the bees with sugar, from sugar cane, was higher showing the predominance of the influence of C_4 plant in the origin of the sugars produced for the bees.

For two samples which pretend to be honeys collected from eucalyptus (sample 10) and citrus (sample 11) we found the $(^2H/^1H)_I$ relationship values 112.2 and 112.0 ppm, respectively. These are very similar to that found for honey produced when the bees are feeding with sucrose solution (111.8 ppm). Probably this sample was adulterated.

Besides the SNIF-NMR could separate sample adulterated when we feed bees with sugar cane, it is not able to distinguish honey sample with different origin like eucalyptus, citrus and wildflower. Train to solve this we use the ^1H NMR spectroscopy.

In figure 2 is showed a honey ^1H NMR spectrum, where can be observed that the majority compounds have been identified as being the glucose and fructose. Moreover, small amounts of oligosaccharides were verified, predominantly sucrose and maltose, and other minority compounds that give proper characteristics to each honey type[18].

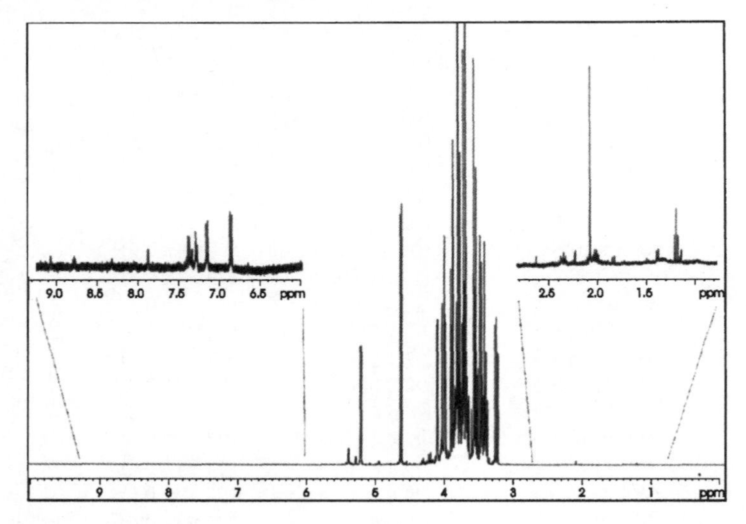

Figure 2 *^1H NMR spectrum of a honey sample*

The samples eucalyptus 10 and citrus 11, were bought as like of eucalyptus and citrus honeys but they presented the ^1H NMR spectra very similar. This is an indication that this samples have unlike characteristics to others, therefore after comparison of all the ^1H NMR spectra for all honeys we do not observed spectra so similar for honeys of different kinds. In figure 3 is showed a comparison between ^1H NMR spectra from citrus (citrus 11 and 7) and eucalyptus (eucalyptus 9 and 10) honeys. The difference observed in this spectra were attributed to three substances, that were identified by gCOSY and gHMBC NMR experiments. Ethanol, a triplet in 1.1 ppm referring to its methyl group; citric acid, two doublets in 2.7 and 2.9 ppm and 5-hydroxymethylfurfuraldehyde (HMF) with signals in the region of 4.5 to 9.7 ppm.

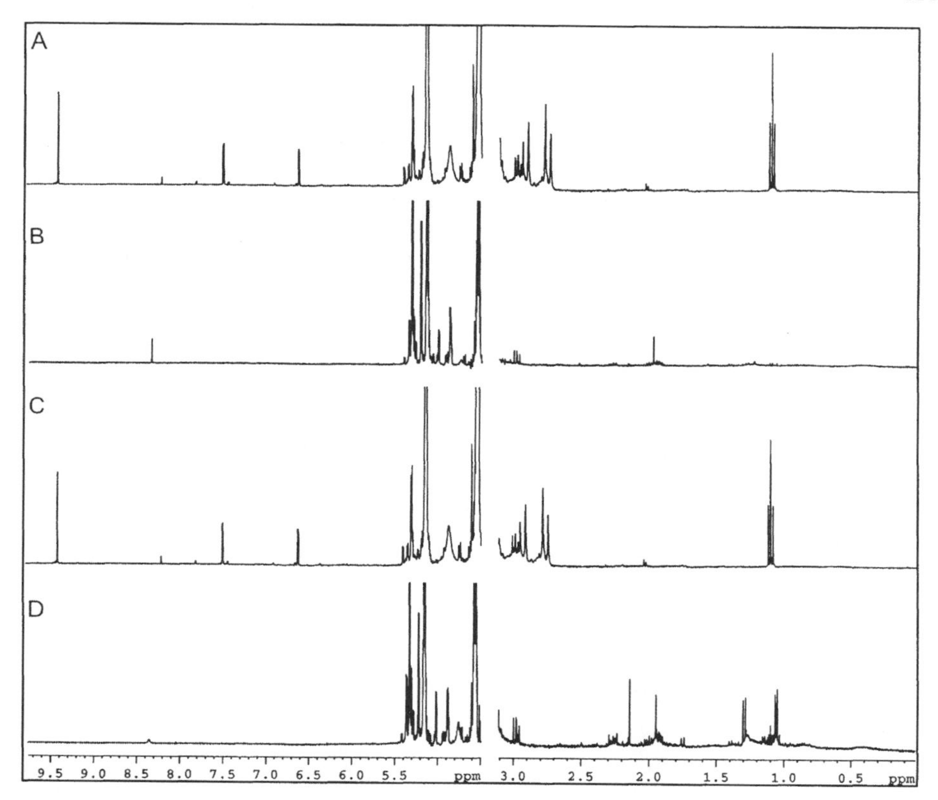

Figure 3 *^{1}H NMR spectra showing the region of 0.0 to 3.1 and 4.5 to 9.7 ppm from honeys (A) citrus 11, (B) citrus 2, (C) eucalyptus 10 and (D) eucalyptus 8*

The presence of ethanol in these honeys (eucalyptus 10 and citrus 11) is an indicative that the fermentation was beginning probably during the storage. Citric acid probably was intentionally added to act as antioxidant, since it was not observed in the ^{1}H NMR spectra for the other citrus honeys; and 5-hydroxymethylfurfuraldehyde (HMF) has been very used as marker in adulteration of the honeys by addition of sucrose, fructose, and other sugars. However, it is also formed by heating the honey when we have a long period product storage and/or it are storage in an inadequate conditions; variations in pH and others factor[19-23]. Therefore the HMF as an indicative of adulteration has been questioned. Honey processing frequently requires heating for both, to reduce viscosity and to prevent crystallisation or fermentation. This results in HMF production, which is formed during acid-catalysed dehydration of hexoses. The presence in honey of simple sugars like glucose and fructose and many acids is a favourable condition for production of this substance[20]. Moreover, in recent years the presence of HMF in foods has been increased a toxicological concerns if the compound and its similar derivatives (5-chloromethyl and 5-sulphidemethylfurfural) have been show to have cytotoxic, genotoxic and tumoral (colon-rectum, hepatic and skin cancers) effects. However, further studies suggest that HMF not offer serious risks to the heath, but the subject still a matter of debate[24].

The application of Hierarchical Cluster Analysis (HCA) in the ^{1}H NMR spectra permitted a discrimination of the citrus and eucalyptus honeys, and moreover separated other honey kinds. The dendrogram obtained is shown in figure 4. The line dotted with a similarity index of 0,452 discriminates the wildflowers, citrus, eucalyptus honey kinds and those that presented anomalous behaviour eucalyptus 10 and citrus 11 samples.

In an analysis more detailed of each group we can observe that the wildflowers, citrus and eucalyptus honeys showed distinct groups, with few exceptions for some wildflowers honeys that were grouped to citrus (wildflowers 2, 19 and 21), probably because they have predominance of nectar collected from citrus plant. We also observe two other groups that are classified like wildflower and citrus. They are assa-peixe (*Vernonia sp*) and sugar-cane (sugar-cane syrupy used to the feed bees). The probably reason for this is that the native assa-peixe area are very small and the bees need to collect nectar from the others kinds of flowers around the area. In the case of sugar-cane, probably it inclusion in the wildflower group should be because it is a more heterogeneous group.

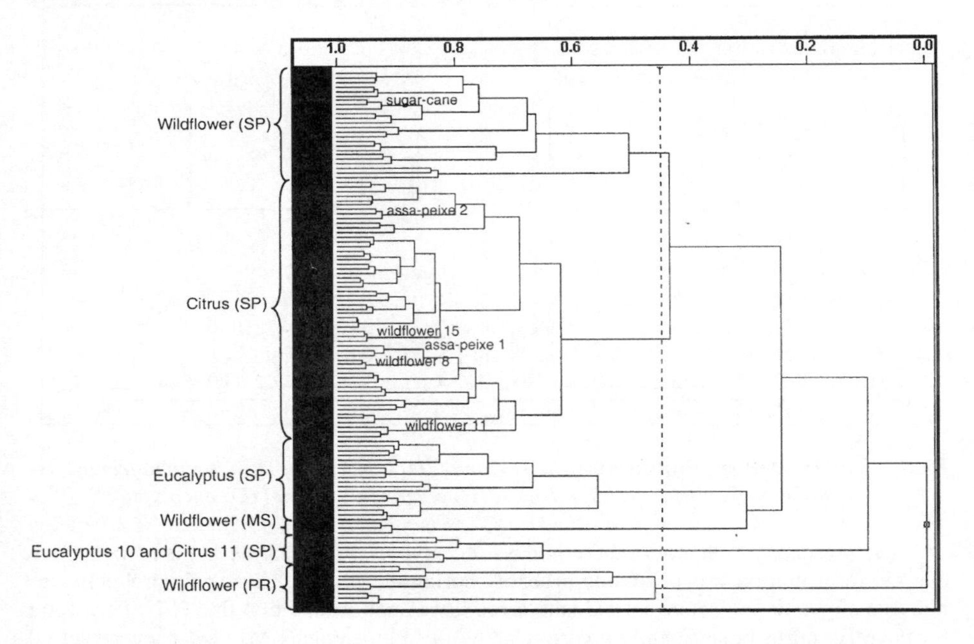

Figure 4 *HCA dendrogram obtained from ¹H NMR spectra from different kinds of honeys*

Another tendency observed was the discrimination regarding the geographical origin, where the three groups located in superior part of dendrogram are the samples originated from São Paulo state (SP), in sequence one sample from Mato Grosso do Sul state (MS) and three from Paraná state (PR).

4 CONCLUSIONS

The SNIF-NMR method can distinguish honeys which were produced from flowers with different biosynthetic origin as C_3 (eucalyptus and citrus) and C_4 (sugar-cane) and chemometric analysis applied to ^1H NMR spectra, for the same samples, can discriminate honeys from eucalyptus and citrus.

Acknowledgements

We are grateful to financial support from CAPES, CNPq, FAPESP and FINEP.

References

1 BRASIL, Secretaria de Comércio Exterior/ Ministério do Desenvolvimento, Indústria e Comércio Exterior, 2004, http://www.midc.gov.br, acessado em 04/10/2004.
2 C. Cordella, I. Moussa, A.C. Martel, N. Sbirrazzuoli, L.L. Cuvelier, *J. Agric. Food Chem.*, 2002, **50**, 1751.
3 BRASIL, Mistério da Agricultura, Pecuária e Abastecimento, Instrução Normativa n° 11 de 20 de outubro de 2000, http//www.agricultura.gov.br, acessado em 05/04/2004.
4 G.J. Martin, C. Guillou, M.L. Martin, M.T. Cabanis, Y. Tep and J. Aerny, *J. Agric. Food Chem.*, 1988, **36**, 316.
5 G. Martin, C. Guillou and Y.L. Martin, *Fruit Processing*, 1995, **5**, 246.
6 A. Hermann, *Eur. Food Res. Technol.,* 2001, **212**, 683.
7 G. Remaud, Y.L. Martin, G.G. Martin, N. Naulet and G.J. Martin, *J. Agric. Food Chem.*, 1997, **45**, 1844.
8 G.G. Martin, R. Wood and G.J. Martin, *J. AOAC Int.*, 1996, **79**, 917.
9 G.G. Martin, V. Hanote, M. Lees and Y.L. Martin, *J. AOAC Int.*, 1996, **79**, 62.
10 C. Vallet, R. Said, C. Rabiller and M.L. Martin, *Bioorg. Chem.,* 1996, **24**, 319.
11 G.J. Padovan, D. Jong, L.P. Rodrigues and J.S. Marchini, *Food Chem.,* 2003, **82**, 633.
12 P. Lindner, E. Bermann and B. Gamarnik, *J. Agric. Food Chem.,* 1996, **44**, 139.
13 G.J. Martin, M.L. Martin, F. Mabon and M.J. Michon, *J. Agric. Food Chem.,* 1983, **31**, 311.
14 G.J. Martin, M.L. Martin and B.L. Zhang, *Plant, Cell and Environment* 1992, **15**, 1037.
15 K.R. Beebe, R.J. Pell and M.B. Seasholtz, *Chemometrics – A Practical Guide*. John Wiley & Sons, New York, 1998.
16 G.J. Martin, X.Y. Sun, C. Guillou and M.L. Martin, *Tetrahedron*, 1985, **41**, 3285.
17 M.L. Martin, G.J. Martin and C. Guillou, *Mikrochim. Acta*, 1991, **II**, 81.
18 C.A.B. Maria and R.F.A. Moreira, *Quím. Nova*, 2003, **26**, 90.
19 B. Fallico. M. Zappalà, A. Arena and A. Verzera, *Food Chem.*, 2004, **85**, 305.
20 M. Zappalà, B. Fallico, A. Arena and A. Verzera, *Food Control*, 2005, **16**, 273.
21 M.L. Sanz, M.D. Castillo, N. Corzo and A. Olano, *J. Agric. Food Chem.*, 2003, **51**, 4278.
22 E.A. Tosi, E. Ré, H. Lucero and L. Bulacio, *Lebensm.-Wiss. u.-Technol.*, 2004, **37**, 669.
23 E.A. Tosi, M. Ciappini, E. Ré and H. Lucero, *Food Chem.*, 2002, **77**, 71.
24 N. Spano, L. Casula, A. Panzanelli, M.I. Pilo, P.C. Piu, R. Scanu, A. Tapparo and G. Sanna, *Talanta*, 2006, **68**, 1390.

THE PRACTICAL ASPECTS OF THE QUANTITATIVE ANALYSIS OF SOLID-LIQUID SYSTEMS USING TD-NMR WITH LOW-FIELD INSTRUMENTS

L. Andrade[1], W. MacNaughtan[1] and I.A. Farhat[2]

1. Division of Food Sciences, The University of Nottingham, Sutton Bonington Campus, Loughborough, LE12 5RD, UK
2. Firmenich S.A., CH-1217 Meyrin, Switzerland

1 INTRODUCTION

This chapter is based on a poster presentation, which attracted significant interest. The decision was taken to develop its content in these proceedings.
Time Domain NMR (TD-NMR) using low field instruments has established itself in many food applications as a technique that combines a wealth of compositional (water content, oil content, solid fat content, etc.) and dynamic (mobility, proton exchange, diffusion, etc.) information with speed of analysis and minimal perturbation to the sample[1,2,3]. However, industrial applications are still limited because of a perceived lack of reliable and robust experimental protocols. The main reason for this is the ineffective information transfer from NMR specialists to quality control analysts in terms of practical aspects that should not be overlooked. This study demonstrates the impact that some of these practical considerations can have on quantitative analysis of the liquid (or solid) content in solid-liquid systems using low-field TD-NMR.
The hardware characteristics of the NMR equipment, such as magnetic field homogeneity, optimum field homogeneity over the sample volume and radiofrequency (RF) coil length[1] as well as practical aspects, such as sample volume / tube filling height and data signal-to-noise ratio (S/N) were investigated.

Of particular interest in this study are:

- The lack of transferability of NMR protocols across equipment due to the reasons above. This is a major limitation facing the industrial implementation of analytical methods based on NMR. The fact that the results may depend on the spectrometer used is mainly due to the unique characteristics of each permanent magnet. In this study, the extent of variance in the NMR signals between spectrometers is demonstrated and addressed.
- The influence of S/N on the quantitative ability of NMR was studied by applying Partial Least Squares Regression (PLSR) to simulated NMR data of samples with different liquid/solid ratios.

Most of the content of this work is "common sense" for any NMR specialist but often not to many users of the technique who are the intended readers of this chapter.

2 METHODS AND RESULTS

2.1 Equipment

Three NMR spectrometers were available for this study and will be referred as spectrometers A, B and C (see Table 1). The aim of this section is to demonstrate the importance of magnet characteristics and not to compare spectrometers.

Table 1 *Description of the 3 spectrometers used in this study.*

Spectrometer	A	B	C
B_0 Frequency (MHz)	20	23	25
Probe internal diameter (mm)	10	10	10
Temperature (°C)	40	25 / 40	25 / 40
Estimated year of manufacturing	1985-1990	1995-1997	2003

2.2 Length of the radiofrequency (RF) coil

In this part, simple protocols to monitor coil length and position relative to the sample tube are described (Figure 1a) as well as the determination of the sampling volume with maximum field homogeneity. These steps are important when intending to implement quantitative analysis on an instrument with unknown configuration.

Determination of coil length:
A small amount of water containing 0.5% $CuSO_4$ (to reduce the proton T_1 and repeat time) leading to a sample height of ~1mm was placed in a flat bottom NMR tube. Free Induction Decay (FID) signals were acquired using spectrometer B at 25°C for different positions of the tube in the sampling area of the probe (moving the tube manually upwards). The results (Figure 1b) indicate that, in the present instrument configuration, the coil is sampling from the very bottom of the tube and that material beyond a height of 11mm is not sampled.

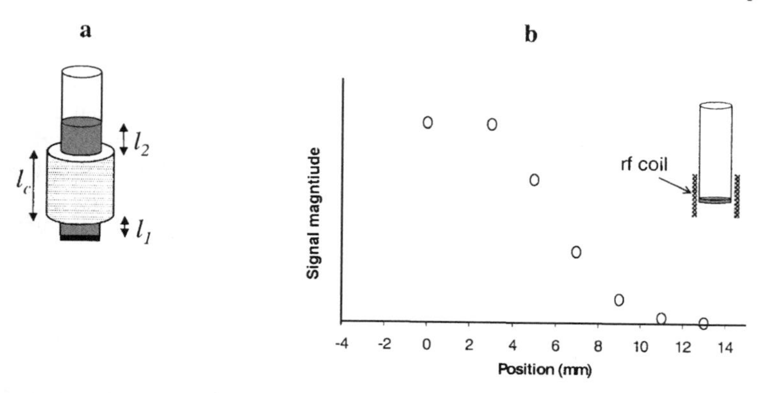

Figure 1 *Determination of the coil position and useful length:*
a: Coil characteristics investigated – coil length, l_c; unsampled tube lengths: bottom, l_1 and top, l_2.
b: Experimental results. Insert: current configuration.

Referring to Figure 1a, in the current configuration: $l_l<0$ and $l_c \geq 11$mm. In fact, the coil is ~20mm but the tube height had been adjusted to ensure that the sample was situated in its centre (Figure 1b, insert).

In addition to the loss of signal intensity, the FIDs acquired on samples where the height of the tube from the bottom position exceeded 5mm decayed more rapidly suggesting a degradation of the homogeneity of the magnetic field.

2.3 Effect of sample volume

Different approaches are adopted during quantitative TD-NMR analysis: under-filling or over-filling the tube (relatively to the coil length / sampling volume). While the effect of tube filling / amount of sample on signal intensity and measurement sensitivity is often accounted, for example, by normalizing to the weight of the sample, its impact on the "quality" of the NMR FID is often overlooked. This effect is illustrated in Figure 2 where FIDs acquired on different amounts of water using spectrometer B (at 25°C) are displayed.

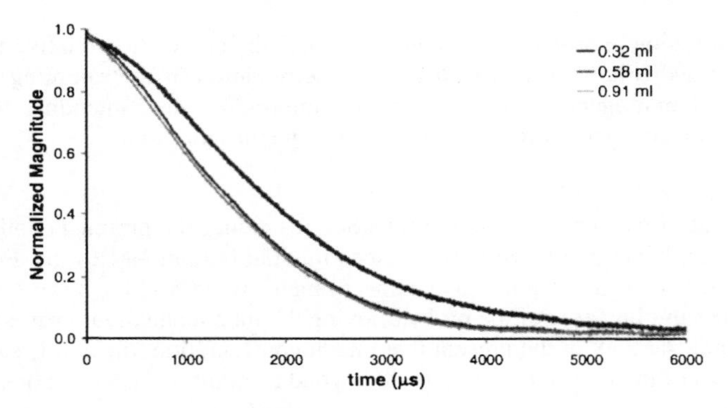

Figure 2 *Normalized FIDs acquired on different volumes of water containing 0.5% $CuSO_4$.*

Increasing the amount of material in the NMR tube resulted in an increased decay rate due to the enhancement of the magnetic field heterogeneities sampled by the larger amount of sample. In this particular instrument configuration, the high homogeneity region was found to be of only ~3mm height or ~0.2ml volume.
Furthermore, it is worth noting that it is usually difficult to optimise the B_0 offset (O_1) in the case of large sample volume of materials containing a significant contribution of slow decaying components. This is reflected by an oscillatory FID magnitude behaviour.

2.4 Transferability of basic TD-NMR measurements between spectrometers

One of the essential challenges for any analytical technique is the requirement of transferring methods between different instruments. This challenge is even more pressing in the case of low field NMR spectrometers based on permanent magnets, as the characteristics of the magnetic field vary significantly between magnets. In order to illustrate this issue and propose a practical solution, measurements were carried out on the three low field spectrometers available (see above), at 40°C, on two types of samples: a solid (crystalline sucrose) and a liquid (sunflower oil). FID and CPMG decays where acquired as a single scan with an optimised receiver gain (signal amplification) setting. Figure 3 clearly demonstrates that, while the FIDs of solid systems, where the decay is dominated by near static dipolar effects, do not significantly vary between instruments, the FIDs of highly mobile systems are extremely sensitive to instrumentation, in particular, the magnetic field homogeneity across the sample.

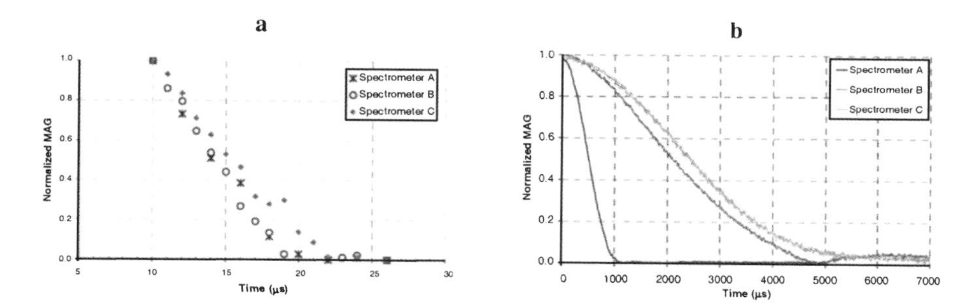

Figure 3 *FID decays of sucrose (a) and sunflower oil (b) acquired using 3 different spectrometers.*

The T_2^* values for the sunflower oil, obtained by fitting a Gaussian line-shape to the FID decays, clearly illustrate this effect – spectrometer A: T_2^*=0.60 ms; B: T_2^*=2.62 ms and C: T_2^*=2.97 ms.

In contrast, and as expected, the CPMG signals acquired on the sunflower oil using the three spectrometers were identical (Figure 4).

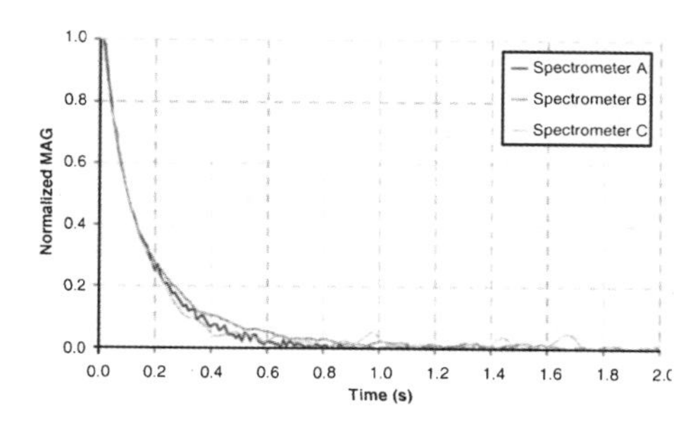

Figure 4 *CPMG decays of sunflower oil acquired using 3 different spectrometers.*

The T_2 values, obtained from the CPMG results fitted to a single exponential decay, were similar at ~144ms ± 3ms for the three spectrometers used.
These expected observations are summarised in Figure 5, which highlights the dependence of the relaxation time of very mobile systems on magnetic field heterogeneity. The apparent relaxation times (T_2^*) determined from the FIDs are much shorter than the actual T_2 value (the limits on its value at high rate of motion are affected by processes such as diffusion and exchange).

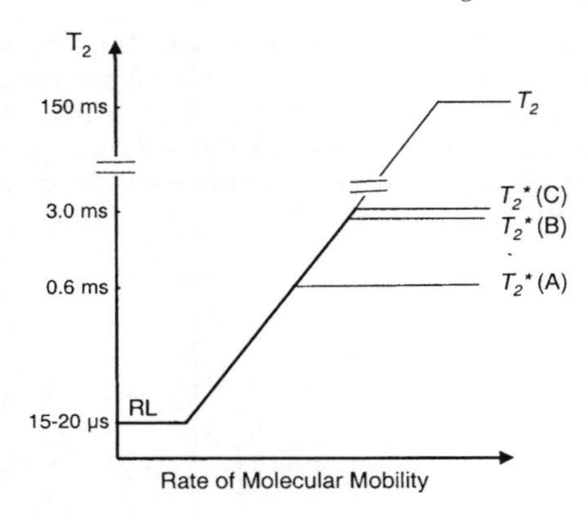

Figure 5 *Summary of the effect of magnetic field heterogeneity (spectrometer A>>B>C) on the measured T_2 in relation to molecular dynamics.*

However, in the analysis of systems containing rigid and mobile components, both FID and CPMG signals are required in order to probe the solid-like (rapidly decaying) and liquid-like (slowly decaying) components, respectively. This can be achieved by combining both pulse sequences, i.e. acquiring FID and CPMG as described in Figure 6. This pulse sequence consists in:

- Acquiring the early part of the FID decay (t_1), typically the first 100μs to sample the solid-like component;
- Waiting for a period of time ($=\tau - t_1$) before applying the train of CPMG 180° pulses (1 or more points can be acquired at the top of each echo).

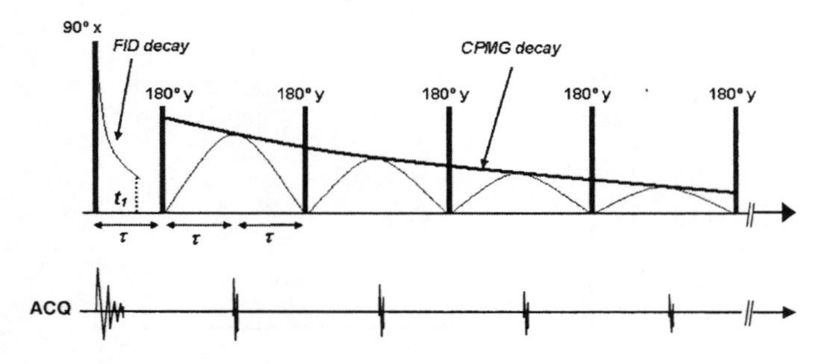

Figure 6 *The FID-CPMG pulse sequence.*

This approach resulted in NMR signals for systems containing different fractions of solid and liquid components, which did not depend on the spectrometer used. An example is shown below for a commercial flavour delivery system containing flavour oil (liquid) encapsulated in a glassy carbohydrate matrix (solid).

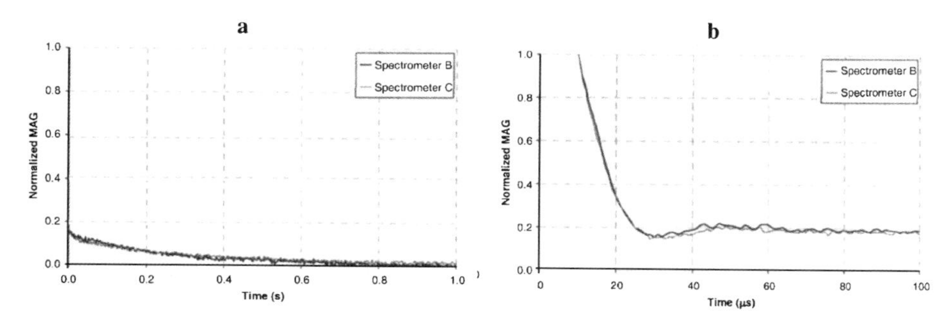

Figure 7 *FID-CPMG decay for a flavour delivery system acquired using spectrometers B and C (a). The FID part of the signal which cannot be clearly distinguished in (a) is shown in (b).*

2.5 Signal-to-noise Ratio (S/N)

Although one can easily understand the impact of the quality of experimental data, in terms of signal-to-noise ratio (S/N), on the ability to obtain quantitative analytical information, this aspect is often overlooked during the development and subsequent implementation of low-field NMR methods in food analysis. In order to highlight the importance of an optimised S/N, FID signals of typical food systems containing a solid and a liquid components were simulated with varying S/N ratios. The FIDs were computer-generated as bi-Gaussian line-shapes (equation 1), illustrating both solid relaxation and highly mobile liquid relaxation [4].

$$A(t) = \sum_i A_i \exp\left(-\frac{t^2}{T_{2i}^2}\right) \quad (1)$$

Where $A(t)$ is the FID magnitude at time t, A_i is the magnitude and T_{2i} is the transverse relaxation time for component i.

The T_2 values used where 20 and 2000μs for the solid and the liquid components, respectively. Data for liquid content ranging between 1.6 and 15% were generated. This range was selected because of its relevance to a wider project on the use of TD-NMR to study flavour delivery systems, which typically contain between 5 and 15% flavour oil.

Variable magnitudes of noise were introduced in the simulated FIDs; the noise data were generated using *Matlab* according to equation 2.

$$Noise = M \times RAND(n, s) \quad (2)$$

Where $RAND(n,s)$ is a randomising function that returns n random numbers (= number of FID points) between 0 and 1 for s samples (number of FIDs) and M is the randomiser magnitude. In this case, n=3000 (FID generated between 0 and 6000μs with a dwell time of 2μs) and s=8 (8 samples with different liquid content).

The value of S/N for each FID was estimated as the total signal magnitude divided by the standard deviation of the noise, which was varied changing the value of M.

Partial Least Squares Regression (PLSR), which is commonly used in quantitative analysis for example, to correlate spectroscopic data with related chemical/physical information[5] was used to assess the quality if the the "calibration" datasets generated with different S/N ratios in providing quantitative information. This was performed using the *Unscrambler 9.2* software (CAMO). Linear PLSR models for the prediction of the liquid content were evaluated on the basis of the Root Mean Square Error of Cross Validation (RMSECV). Generally, a low value for RMSECV is required and models that include a low number of PLS components in the description of the data are considered more reliable.

As expected, the prediction error (RMSECV) of the PLSR models decreased dramatically as the data S/N ratio increased (Figure 8) – from an unacceptable level of RMSECV of ~5.3% at S/N=3 down to ~0.2% at S/N=130 and to less than 0.05% at S/N=326. Further improvement to S/N did not bring noticeable benefit. (Note that the range of liquid contents was 1.6-15%.)

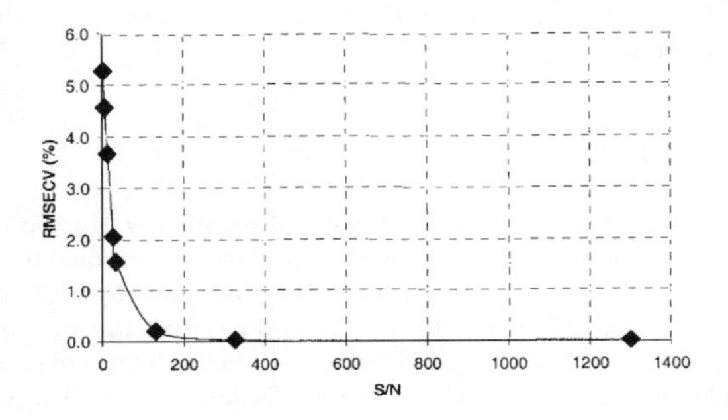

Figure 8 *Dependence of prediction error (RMSECV) on signal-to-noise ratio (S/N).*

To emphasize these observations, the plots of the predicted liquid contents versus the measured ones and the statistical results for S/N=3 and S/N=130 are presented in Figure 9. At low S/N (Figure 9a), data computation resulted in an unsuitable model to predict the liquid content and all 8 samples were predicted to have a liquid content value between 7% and 9%. Increasing S/N to 130 improved greatly the performance of the model (Figure 9b).

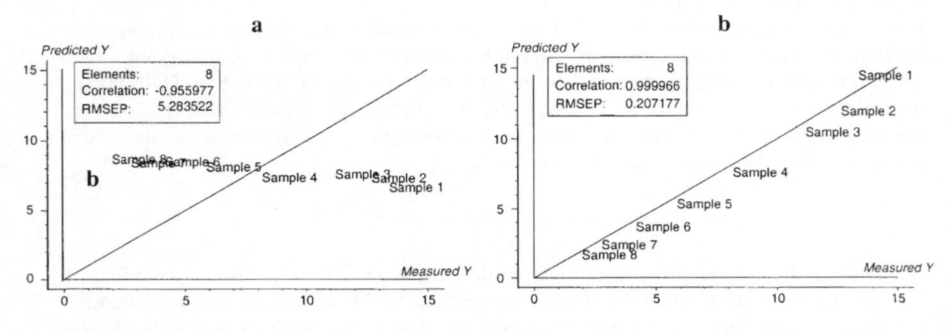

Figure 9 *Predicted vs. Measured Liquid Content Values plots and statistical results for S/N=3 (a) and 130 (b).*

The signal-to-noise ratio can be improved through one or several of the following approaches:
- Increasing the size of the sample – however, this must be considered at the expense of the quality of the NMR signal (see section 2.3);
- Optimising signal amplification;
- Increasing the number of scans: the signal-to-noise increases with the square root of the number of co-added scans. The number of scans selected is usually a compromise between S/N requirements and experimental time implications.

The lower limit of S/N depends on the desired value for the final prediction error.

3 CONCLUSIONS

While TD-NMR using low field instruments is widely established in food analysis[6,7,8] many basic aspects of the technique are often overlooked. These can be vital for a successful implementation of robust and reliable protocols in industrial applications. The principal reason for this oversight is the fact that NMR protocols are often implemented by quality control analysts with limited NMR training (this is not surprising as most food sciences courses lack such training). This chapter attempted to illustrate the most common shortfalls. Other important aspects, which have not been covered, include (the list is not exhaustive):

(i) Insuring maximum relaxation between scans: this is particularly important for systems containing components with large differences in relaxation characteristics, in which some components can be unintentionally saturated through short delays between scans;
(ii) Pulse/acquisition phase cycling to eliminate unwanted signal contributions such as those resulting from imbalances in spectrometer hardware, coherent noise, artefacts generated by multiple pulses, etc.[9];
(iii) Other: temperature equilibration, changes in water content, etc.

The chapter described a simple approach for the implementation of a protocol (FID-CPMG) allowing the recording of NMR data for systems containing both solid-like and liquid-like components, which does not depend on the spectrometer used and hence can be transferred between instruments.

References

1. W. Kemp, NMR in Chemistry – A Multinuclear Introduction, *The Macmillan Press LTD*, 1986.
2. J. McCarten, Fat Analysis in Margarine Base Products, *Bruker-Physik AG*, Karisruhe, *Bruker Minispec PC100*, Typical Applications in Food Industry.
3. K. van Putte, Pulsed NMR as a routine method in the fat and margarine industry I, *Unilever Research*, Vlaardingen, The Netherlands, *Bruker Minispec PC100*, Typical Applications in Food Industry.
4. R. Partanen, V. Marie, W. MacNaughtan, P. Forssell, & Imad Farhat, H-1 NMR study of amylose films plasticised by glycerol and water, *Carbohydrate Polymers*, *56*, 147-155, 2004.
5. E. Micklander, Quantitative, Qualitative and Exploratory Analysis of Food Using Spectroscopy and Chemometrics, PhD thesis, *KVL*, Denmark, 2004.

6. P. Lambelet, Comparison of NMR and DSC Methods for Determining Solid Content of Fats – Application to Milk Fat and its Fractions, *Nestlé Products Technical Assistance Co.*, Research Department, Switzerland, 1982.
7. G. H. Sorland, P. M. Larsen, F. Lundby, A. Rudi, & Thierry Guiheneuf, Determination of total fat and moisture content in meat using low field NMR, *Meat Science*, 66: 543-550, 2004.
8. K. G. Hollingsworth, A. J. Sederman, C. Buckley, L. F. Gladden, & M. L. Johns, Fast emulsion droplet sizing using NMR self-diffusion measurements, *Journal of Colloid and Interface Science*, 274: 244-250, 2004.
9. F. Sauriol, Phase Cycling in 1D NMR, *http://www.chem.queensu.ca/facilities/NMR/nmr/webcourse/*

Food Processing

INFLUENCE OF GRAIN STRUCTURAL COMPONENTS ON THE DRYING OF WHEAT: A MAGNETIC RESONANCE IMAGING STUDY

P.K. Ghosh[1], D.S. Jayas[1], M.L.H. Gruwel[2,3] and N.D.G. White[1,4,5]

[1]Department of Biosystems Engineering, University of Manitoba, Winnipeg, Manitoba, R3T 5V6, Canada
[2]NRC Institute for Biodiagnostics, 435 Ellice Avenue, Winnipeg, Manitoba, R3B 1Y6, Canada
[3]Department of Chemistry, University of Manitoba, Winnipeg, Manitoba, R3T 2N2, Canada
[4]Cereal Research Centre, Agriculture and Agri-Food Canada, 195 Dafoe Road, Winnipeg, Manitoba, R3T 2M9, Canada
[5]Department of Entomology, University of Manitoba, Winnipeg, Manitoba, R3T 2N2, Canada

1 INTRODUCTION

Wheat is a staple grain, which is consumed in various forms around the world. Global wheat production was about 628 million tonnes (Mt) in the year 2005 from a total harvested area of about 216 Mha.[1] An increased wheat production of about 85 Mt over the last decade indicates an increasing demand due to the world's growing population. Therefore, safer and proper storage of wheat has become a major concern for farmers or grain producers. The recommended harvesting moisture content for wheat is usually 18-20% wb (wet mass basis), whereas the safe storage moisture content is recommended to be 14% wb for short-term (6-12 mo) and 13% wb for long-term storage (over 1 y).[2] Drying of wheat is globally practiced to reduce moisture of grain for safe storage. Wheat is usually dried using near-ambient air or hot air with a condition that wheat is not heated above 60°C to maintain its milling and baking qualities (e.g., loaf volume).

A kernel of wheat consists of three major components, the endosperm (82-85% by weight), the germ or embryo (2-3% by weight), and the pericarp (13-15% by weight). Detailed information on a wheat kernel structure can be obtained from several excellent sources.[2-5] The complex anatomy of a wheat kernel makes drying a complicated process of simultaneous heat and mass transfer.

Several researchers have studied the internal moisture distribution as well as moisture transfer pattern inside a grain kernel during drying, cooking, boiling, and soaking using magnetic resonance imaging (MRI) technology.[6-21] Results, particularly from grain drying studies, revealed that moisture content distribution is non-uniform in a grain at the beginning of drying and that the moisture removal from the grains is non-uniform during drying. Ghosh et al.[11,12] compared and explained the influence of the structural components of a Canadian variety wheat kernel on the internal moisture movement during drying using MRI with three physically different types of wheat kernels: intact or whole kernels, mechanically scarified kernels with incisions in the pericarp, and kernels with the embryo removed. The moisture transfer from all three different wheat kernels was anisotropic and non-uniform during drying. For the intact wheat kernels, moisture appeared to move through the embryo-end from the endosperm whereas the pericarp dried faster at the initial stages of drying and presumably acted as an insulator. The mechanically-scarified kernels lost water relatively fast from the incision part of the pericarp at the initial stages of drying. Water was then moved from the endosperm into the interior of the embryo region through the scutellum epithelium. During drying, the path of water movement in embryo-removed kernels was shown to follow the capillaries of the endosperm towards the aleurone layer of the pericarp. Overall, it has been found that the embryo (germ) plays a significant role in moisture migration during drying. However, no attempt has been made to study the effect of temperature on the grain parts during drying. Knowledge from this study is expected to be incorporated in future drying model developments. Therefore, the objective of this investigation was to compare the effect of temperature on the MR signals and on the drying curves computed for the three physically different kinds of wheat kernels.

2 METHOD AND RESULTS

Canada Western Red Spring (CWRS) wheat (*Triticum aestivum* L., cv. 'A.C. Barrie') containing approximately 15.7±0.03% db (dry basis) moisture content was used in this study. All MRI experiments were performed on a 11.7 T (500 MHz) Magnex (Magnex Scientific Ltd., Yarnton, UK) super-conducting vertical-bore magnet and an in-line variable temperature and flow controller unit (Bruker BVT-1000, Milton, ON) attached to the custom built MRI probe (Helmholtz configuration, 7 mm internal diameter). Prior to MRI, wheat kernels were preconditioned to a known moisture level (greater than 25% db) to acquire bright MR images that helped in studying the intrinsic details. For every MRI experiment, individual kernels were placed into the imaging probe and placed into the bore of the vertical MRI magnet. Drying was started following a short setup period, and continued for 4 h at 30, 40, and 50°C at a constant nitrogen gas flow (~0.23 m/s). Moisture movement within and out of a wheat kernel during drying was observed under the following conditions: initial average moisture contents: 60.3±0.8% db (intact or scarified kernels) and 49.7±3.2% db (embryo-removed kernels); three types of wheat kernels: intact, mechanically-scarified, and embryo-removed; and drying temperatures: 30, 40 and 50°C with an accuracy of ± 1°C. Images were obtained using a two-dimensional, multi-slice Hahn spin-echo pulse sequence.[22] The repetition time (TR) of 200 ms and the echo time (TE) of 3.375 ms were set to produce the acceptable image contrast. A total of eight slices, each of 0.5 mm thickness, were obtained from each wheat kernel with an image resolution of 100 μm × 200 μm × 500 μm. Image acquisition time was 10 min 18 s for six scans, acquired for signal averaging. The MR data matrix size was increased by zero-filling. A particular transverse slice out of 8 slices was selected to best illustrate the moisture distribution in the kernel with the greatest anatomical detail for the analysis of drying as a function of time.

To observe the influence of the kernel components on the drying process, we used samples of intact kernels (with all three components: pericarp, embryo, and endosperm), mechanically scarified kernels by making incisions in the pericarp, and kernels with the embryo removed. The intact kernels were imbibed for moisture equilibration before starting each experiment. The incisions on the intact kernels were made after the moisture equilibration to prepare the scarified kernels. However, the embryo was removed from the intact kernels prior-to the moisture equilibration. This was done to reduce possible loss or redistribution of soluble materials of the imbibed kernels and thus to achieve more accurate results.[3] The embryo-removed kernels were at lower moisture content (about 50% db) than the intact or the scarified kernels (about 60% db). We were interested in comparing relative drying rates for individual kernel components, therefore drying experiments were replicated three times at each temperature with similar sized wheat kernels selected by visual inspection. Results analyzed in this article are the average normalized image intensities of three replicates.

All the reported image intensities were derived by the following equation to eliminate the effect of the noise on the analysis of drying curves[23]:

$$I_{avg} = \frac{I_{kernel} - I_{noise}}{I_{noise}} \tag{1}$$

where by definition:

$$I_i = \frac{1}{A_{I_i}} \iint_{A_{I_i}} I\ (x, y).\ dx\ dy \tag{2}$$

where, i is the kernel or noise area; I is the pixel intensity at coordinate (x,y); and A_{I_i} is the average area. All the MRI images were processed and analyzed by using various image processing software (Marevisi v. 7.1, ImageJ, and Matlab v. 7 R14). Data were analyzed using the statistical procedures in Microsoft® Excel (v. SP-2, 2002) and SigmaPlot v. 6.0 (SPSS Inc., Chicago, IL). A detailed methodology is given in Ghosh et al.[11,12]

Figure 1 shows representative MR images for an intact kernel, a mechanically scarified kernel, and an embryo-removed kernel during drying at 30°C as a function of drying time. Images in each row were scaled primarily based on the minimum and maximum pixel values in the first image at the beginning of drying. The signal-to-noise ratio (SNR) of the first image (just before drying) in serial MR images were: intact: 31.6, mechanically scarified: 21.7, and embryo-removed: 13.0. The SNR was calculated as the ratio of average pixel intensity of an irregular polygon shaped region of interest (ROI) in the whole grain to the average pixel intensity of the same size area in a region of noise only. As drying proceeds, water moves out from the kernel, the moisture content drops, as does the signal intensity, and therefore the signal-to-noise ratio decreases. The MR images show the variation in the internal distribution of water prior-to and during drying. The pericarp hardly showed any signal intensity at the onset of the drying process since moisture moved out from this part quickly. Water was concentrated mainly in the embryo region and the embryo signal intensity remained high, even after 4 h of drying at all three temperatures. Previously measured[12] higher T_2 values of the embryo region, in combination with the relatively high signal amplitude indicate the presence of mobile ("free") water in the embryo. The lowest T_2 values were observed in the endosperm region and, in combination with the low signal amplitude, indicated that water interacts more strongly with the cellulose matrix.

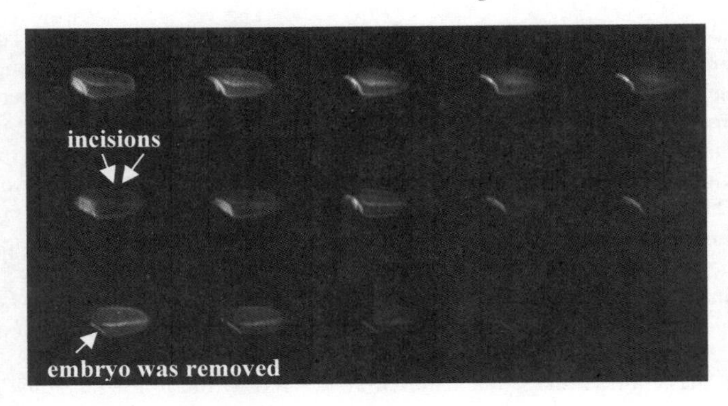

Figure 1 *2D MR image series of intact, mechanically scarified and embryo-removed wheat kernels dried at 30°C after every 1 h from the beginning of drying*

A representative drying curve of the ratio of the averaged local normalized image intensity to the initial normalized image intensity (I/I_o) of an intact wheat kernel at three different temperatures is shown in Figure 2. Although similar sized wheat kernels were selected by visual inspection for the drying experiments, the homogeneity of the grains might differ. Therefore, the mean data were considered at each drying time and the data were analyzed in terms of ratios. The influence of the drying temperature on the overall drying of the wheat kernels is clearly evident from these curves, which indicate faster drying at higher temperatures. In addition, with the exception of the embryo-removed kernel (endosperm only), these drying curves did not indicate a single exponential relationship between the image intensity ratio and the drying time. This indicated that water reduction is dependent not only on the overall water content in the kernels but also on the structural composition of the wheat kernel. An in-depth analysis of influence of structural components on the wheat drying is reported elsewhere.[12]

Semi-logarithmic plots of the intensity ratio of the intact and the mechanically scarified kernels with their major components (embryo and endosperm) at various temperatures versus drying time are shown in Figures 3 and 4. These graphs were made to estimate the differences in the drying rates of different wheat components during drying. The drying rate was increased in general from 30°C to 50°C, but the overall drying rate was much lower in the embryo region than that of the endosperm region for both kinds of kernels. Furthermore, the drying rates for the embryo region became much slower after the first two hours of drying than the endosperm region since the embryo has higher moisture content than the endosperm. These particular shapes of the drying curves are evident at the later stage of drying for hygroscopic capillary-porous materials.

The effect of temperature on the drying curves of individual grain components was analyzed by normalizing the semi-logarithmic plots of intact, mechanically scarified and the embryo-removed wheat kernels. These curves were normalized with respect to 30°C at each drying time after multiplying with a scale factor (k) using a least square method of third-order polynomial regression analysis (Table 1). A representative normalized drying curve for a mechanically scarified kernel is shown in Figure 5. Semi-logarithmic plots of the changing scaling factors with the inverse temperature (K^{-1}) produced an Arrhenius type relationship of the following linearized form to determine the activation energy of the water removal process inside the grain components during drying:

$$ln\ k = -\frac{E_a}{R}\left[\frac{1}{T}\right] + ln\ z \qquad (3)$$

where, a plot of *ln k* (*k* being the scale factor) versus *1/T* (K^{-1}) produce a straight line with a slope of $-E_a/R$ in which E_a is the activation energy, $kJ \cdot mol^{-1}$ and R is the Universal gas constant, $8.314 \times 10^{-3}\ kJ \cdot mol^{-1} \cdot K^{-1}$; z is a constant.

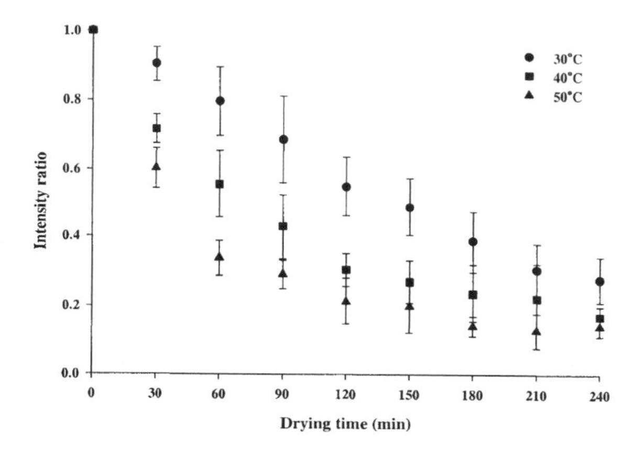

Figure 2 *Drying curves of an intact wheat kernel at different temperature*

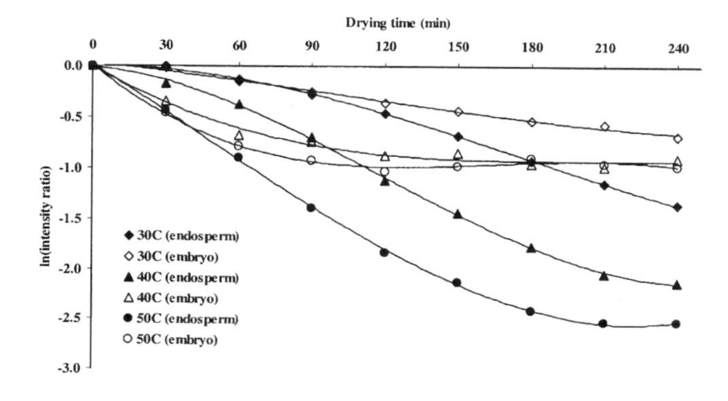

Figure 3 *Drying curves of the intact kernel components at different temperature*

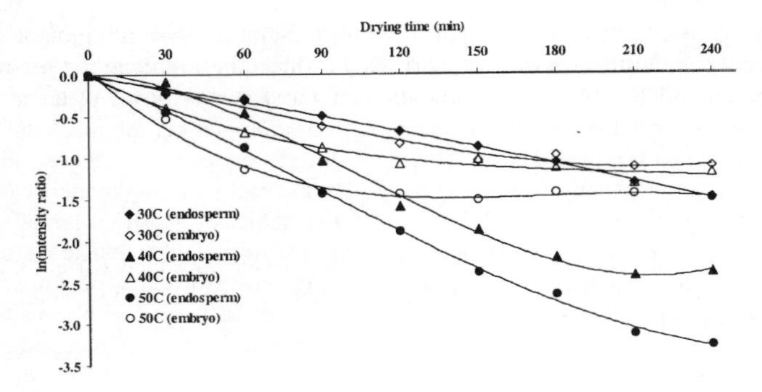

Figure 4 *Drying curves of the mechanically scarified kernel components at different temperature*

Table 1 *Scale factors (k) used to normalize all the drying curves with respect to 30°C*

Temperature (°C)	*Intact wheat*			*Mechanically scarified wheat*			*Embryo-removed wheat*
	Endosperm	*Embryo*	*Whole*	*Endosperm*	*Embryo*	*Whole*	
30	1	1	1	1	1	1	1
40	0.235	0.138	0.275	1.286	0.486	0.711	0.570
50	0.072	0.098	0.176	0.388	0.309	0.549	0.170

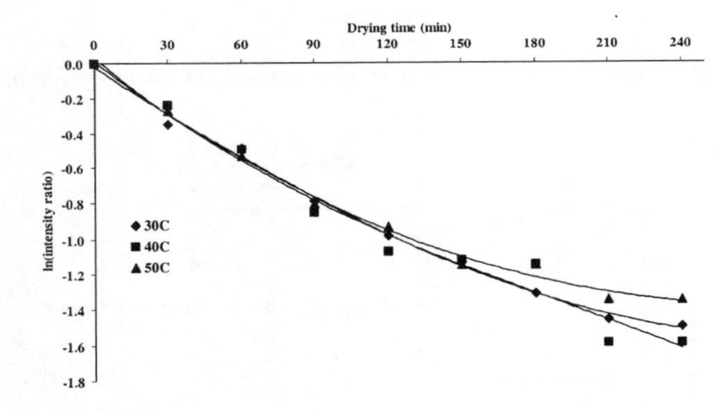

Figure 5 *Normalized drying curves of mechanically scarified kernel (whole kernel) at different temperatures; normalization was done with respect to 30°C*

Figure 6 shows a plot of the natural log of the scale factors of different kinds of wheat kernels along with their structural components for drying at different temperatures versus the inverse of the temperature in K^{-1}. A strong correlation of the water removal process to the drying temperature was observed in case of the endosperm of an intact kernel, mechanically scarified kernel and an embryo-removed kernel. However, for the endosperm the activation energy of the water removal process was the highest in an intact kernel (activation energy of 109 $kJ \cdot mol^{-1}$) compared to an embryo-removed kernel and a mechanically scarified kernel, respectively. In case of the scarified kernel, the pericarp started offering a water resistance due to closely adhered thick-walled cells after initial stages of drying. Therefore, the moisture removal from this part became slower. However, the rate of decrease of factor k was slower after 40°C for the embryo in both the intact and mechanically scarified kernels, which indicated that the drying did not depend on temperatures above 40°C.[12] Furthermore, the activation energy of the water removal process in the embryo was higher in an intact kernel than that in a scarified kernel (activation energy 49 $kJ \cdot mol^{-1}$). Since water was released relatively faster from the scarified regions of the endosperm-pericarp boundary at the initial stages of drying, less "free" water was available in the embryo in the case of the scarified kernels. However, this was not the case in an intact kernel since more water was available in the embryo due to a high osmotic pressure gradient. This might be the reason why an overall activation energy of the scarified kernel (25 $kJ \cdot mol^{-1}$) was less than that of an intact kernel. Ishida et al.[19] reported an overall activation energy change of 63 $kJ \cdot mol^{-1}$ during the drying of a freshly harvested Japanese variety intact rice kernel (20% wb) at a temperature range of 40 to 60°C.

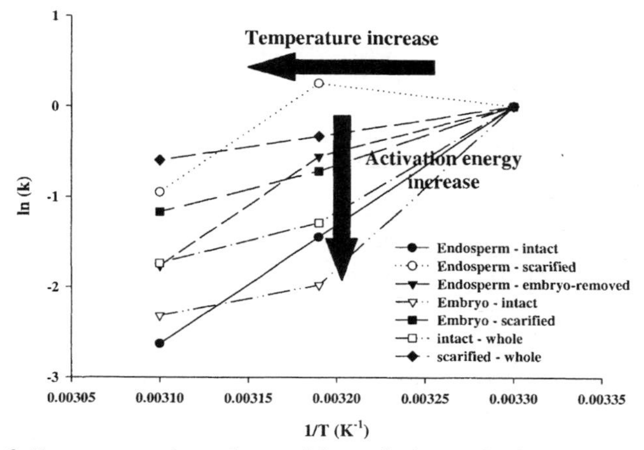

Figure 6 *Temperature dependency of the scale factor (k); k was used to normalize the curves with respect to 30°C*

Thus, the current investigation reveals that the individual components of a wheat kernel play a significant role in its drying process. Results from this study would be beneficial to the development of appropriate drying models considering an entire domain of a wheat kernel comprising of two major sub-domains (embryo and endosperm) with difference in their moisture content values.

3 CONCLUSION

The MRI technique was a useful method to study the influence of the wheat kernel components on drying. Different physical conditions of a wheat kernel helped to analyze the actual rate of drying in the components of wheat. Internal moisture transfer was visualized and drying rate was studied using the MR signal intensity ratio and activation energy of individual components was evaluated. A proper synthesis of the results from these analyses will lead to development of mathematical models to follow the moisture distribution inside a grain kernel as a function of drying time and initial moisture content.

4 ACKNOWLEDGEMENTS

The authors thank the Natural Sciences and Engineering Research Council of Canada, University of Manitoba Graduate Fellowship Committee, and Canada Research Chairs program for funding this study and presentation of this paper.

References

1 FAOSTAT, http://faostat.fao.org/faostat/default.jsp, 2006.
2 D. B. Brooker, F.W. Bakker-Arkema and C.W. Hall, *Drying and storage of grains and oilseeds*, Van Nostrand Reinhold, New York, NY, 1992.
3 M.M. MacMasters, D. Bradbury, J.J.C. Hinton, 'Microscopic structure and composition of the wheat kernel' in *Wheat – chemistry and technology*, ed., I. Hlynka, Amercian Association of Cereal Chemists, Inc, St. Paul, MI, 1964, pp. 55-110.
4 Y. Pomeranz and D.B. Bechtel, 'Structure of cereal grains as related to end-use properties' in *Cereals'78: Better nutrition for the world's millions*, Sixth International Cereal and Bread Congress, American Association of Cereal Chemists, Inc, St. Paul, MI, 1978.
5 A.D. Evers and D.B. Bechtel, 'Microscopic structure of the wheat grain' in *Wheat – chemistry and technology*, ed., Y. Pomeranz, Amercian Association of Cereal Chemists, Inc, St. Paul, MI, 1988, Vol. I, pp. 47-96.
6 C.F. Jenner, Y. Xia, C.D. Eccles and P.T. Callaghan, *Nature*, 1988, **336**, 399.
7 A.G.F. Stapley, M.H. Thomas, L.F. Gladden and P.J. Fryer, *Intl. J. Fd. Sci. and Technol.*, 1997, **32**, 355.
8 H.P. Song, S.R. Delwiche and M.J. Line, *J. Cereal Sci.*, 1998, **27**, 191.
9 M. Fukuoka, T. Mihori and H. Watanabe, *J. Fd. Sci.*, 2000, **65**, 1343.
10 M. Turhan, S. Gunasekaran and B.P. Lamsal, *Drying Technol.*, 2001, **19**, 333.
11 P.K. Ghosh, D.S. Jayas, M.L.H. Gruwel and N.D.G. White, *Canadian Biosys. Engng.*, 2006a, **48**, 7.13.
12 P.K. Ghosh, D.S. Jayas, M.L.H. Gruwel and N.D.G. White, *Trans. ASABE*, 2006b, **49**, in press.
13 H. Song and J.B. Litchfield, *Trans. ASAE,* 1990, **33**, 1286.
14 R. Ruan, J.B. Litchfield, *Cereal Chem.*, 1992, **69**, 13.
15 R. Ruan, J.B. Litchfield and S.R. Eckhoff, *Cereal Chem.*, 1992, **69**, 600.
16 H.P. Song, J.B. Litchfield and H.D. Morris, *J. Agric. Engng. Res.,* 1992, **53**, 51.

17 A.J. Kovács and M. Neményi, *Mag. Res. Imaging*, 1999, **17**, 1077.
18 S. Takeuchi, M. Fukuoka, Y. Gomi, M. Maeda and H. Watanabe, *J. Fd. Engng.*, 1997, **33**, 181.
19 N. Ishida, S. Naito and H. Kano, 2004, **22**, 871.
20 M.L.H. Gruwel, X.S. Yin, M.J. Edney, S.W. Schroeder, A.W. MacGregor and S. Abrams, *J. Agric. and Fd. Chem.*, 2002, **50**, 667.
21 S.M. Glidewell, *J. Cereal Sci.*, 2006, **43**, 70.
22 E.L. Hahn, *Physical Rev.,* 1950, **80**, 580.
23 P. Bernada, S. Stenström and S. Månsson, *J. Pulp and Paper Sci.*, 1998, **24**, 373.

DYNAMIC VISUALISATION OF STRUCTURAL CHANGES IN CEREAL MATERIALS UNDER HIGH-MOISTURE CONDITIONS USING 3D MRI AND XRT

W.P. Weglarz[1], G.J.W. Goudappel, G. van Dalen, H. Blonk and J.P.M. van Duynhoven

Unilever Food and Health Research Institute, Unilever R&D, Vlaardingen, The Netherlands
[1]On leave from the Institute of Nuclear Physics, Krakow, Poland

1 INTRODUCTION

Ingress of bulk water into porous cereal materials causes hydration of the solid matrix through saturation and swelling. Maintaining crispiness under high-moisture conditions for a certain time is an important feature in terms of consumer appreciation. Normally, macroscopic techniques are used to evaluate crispiness, however these does not provide information about mechanisms of internal structure changes.
Magnetic Resonance Imaging (MRI), which is well suited for non-invasive imaging of water migration in tissue, has found a broad range of applications in different areas, including food sciences [1, 2]. Appropriate choice of the MRI measurement methods allows for the real time visualisation of the effect of hydration in terms of rate and space distribution of structural changes. X-ray microtomography (XRT) on the other hand allows for high resolution 3D visualisation and characterisation (e.g. porosity assessment) of the non-hydrated material [3].
The aim of this work was to use MRI to visualise and quantify in real-time the structural changes of the carbohydrate matrix (i.e. swelling) caused by ingress of water and correlate these with structural information from XRT. For this purpose, prototype food systems – soup inclusions with different internal structure were used.

2 MATERIALS AND METHODS

2.1 Samples

Carbohydrate (wheat) based soup inclusions with different internal structure were produced locally. For some measurements samples were coated with plant oil in order to retard water migration.

2.2 MRI methodology

2.2.1 *Properties of NMR signal of hydrated carbohydrate-based matrix*

MR images are based on NMR signals sampled in presence of magnetic field gradients in order to obtain spatial resolution. Properties of the NMR signal from the sample (i.e. proton density, T_2, T_1) as well as signal sampling algorithm (MRI measurement method) define sensitivity and thus image contrast and temporal and spatial resolution. Fig. 1 shows NMR time domain signals (T_2 decays) from moderately hydrated carbohydrate-based solid matrix. It is composed of a short - "solid" component disappearing after 50 - 100 μs depending on the sample, long - "liquid" component with T_2 above ~2ms, and an intermediate component with T_2 in the range roughly between 0.1 ms and 2 ms. Solid component originates from the protons in carbohydrate, protein and solid fat macromolecules constituting the matrix. Intermediate component in most cases describes water at low or moderate hydration level (Aw < ~0.9), while the liquid component corresponds to water that is not associated to the solid matrix and/or to oil usually present in carbohydrate-based food (Fig. 1).

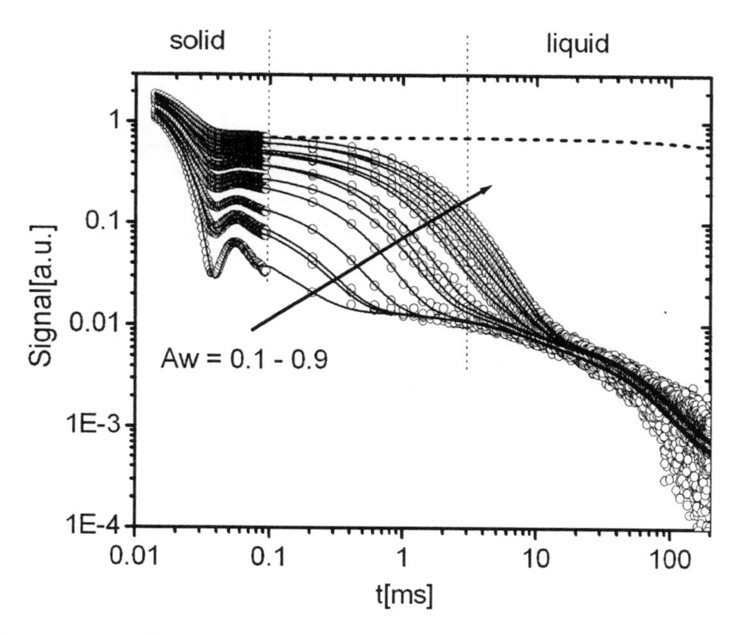

Figure 1 *NMR time domain signal (T_2 decay) from moderately hydrated carbohydrate based matrix. Signal from bulk water shown for comparison (dotted line).*

2.2.2 *Fast Spin Echo Imaging Method (FSE, TSE, RARE)*

In Fast Spin Echo Imaging (also known as Turbo Spin Echo or RARE) a CPMG train of echoes is used to obtain an image. For samples containing "liquid" components with long T_2 this method can be very time- and space- resolution efficient. Single image is produced from signals collected from multiple echoes, measured at different echo time. The possibility to adjust the number of echoes (RARE factor - RF) and the phase encoding scheme allows for different variants of the method for enhancing or suppressing signal with different T_2 (for example see Fig. 2). If the measured sample contains "liquid" regions with different T_2, like bulk water or a water-saturated porous matrix, this will result in a complicated dependence of the obtained image contrast and method setup. FSE is well suited for dynamic measurements of the hydration of carbohydrate matrices. However, if signal from a water saturated matrix (T_2 in order of tens of ms) need to be enhanced in the image, temporal or/and spatial resolution has to be compromised due to a limiting RARE factor. Suppressing of the signal from bulk water can be achieved by applying short repetition time (TR) as compared to bulk water T_1, which typically is in order of 1 s.

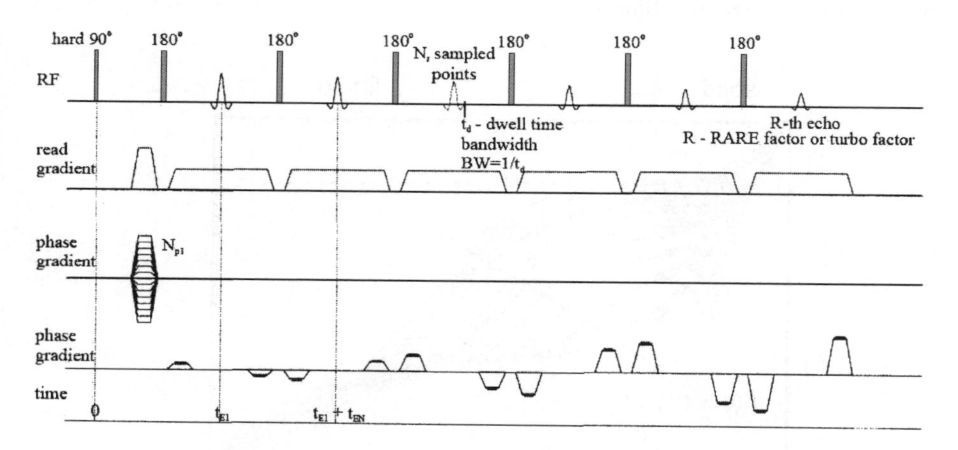

Figure 2 *Centered-out 3D RARE imaging method.*

2.3 X-ray microtomography

XRT can probe the 3D microstructure of samples non-invasively up to a few millimetres across with an axial and lateral resolution down to a few micrometers. The contrast in XRT images is based on the difference in absorption of X-rays by the constituents of the sample (e.g. fat, carbohydrate and air). In cereal materials the contrast is mainly caused by a difference in density. Samples were imaged using a Skyscan 1072 desktop XRT system (Belgium, http://www.skyscan.be). X-rays are generated by a microfocus X-ray tube (10 µm focal spot size) with tungsten anode generating an energy ranging from 10 to 100 keV (peak at 80 kV). The transmission of the conical X-ray beam through the sample is recorded by a CCD camera with 1024 x 1024 pixels. XRT produces two-dimensional images of projections of the sample. A set of flat cross sections was obtained after tomographical reconstruction of images acquired under different rotations over 180 degrees with a step size of 0.45 degrees. The acquisition time for one projection was 2.8 s resulting in a total acquisition and read-out time of about 40 min. Samples were imaged using plastic cylindrical sample holder with an inner diameter of 11.2 mm. For visualisation in 3-D space, isosurface rendering was used (Amira 4.1 from Mercury Computer Systems). This was mainly done by segmentation using thresholding followed by surface generation with constrained smoothing. The surface generation module of the Amira software computes a triangular approximation of the interfaces between the segmented sections.

2 SOUP INCLUSIONS HYDRATION

2.1 Dry sample structure characterisation

The porous structure of soup inclusions was characterised by XRT. 3D images with resolution of 15 μm were obtained, showing clearly differences in internal structure (Fig. 3). In addition, the distribution of the fat coating was assessed based on differences in X-ray attenuation between carbohydrate matrix and fat coating (Fig. 4).

Figure 3 *2D cross-sections through 3D XRT images of soup inclusions with different internal structures.*

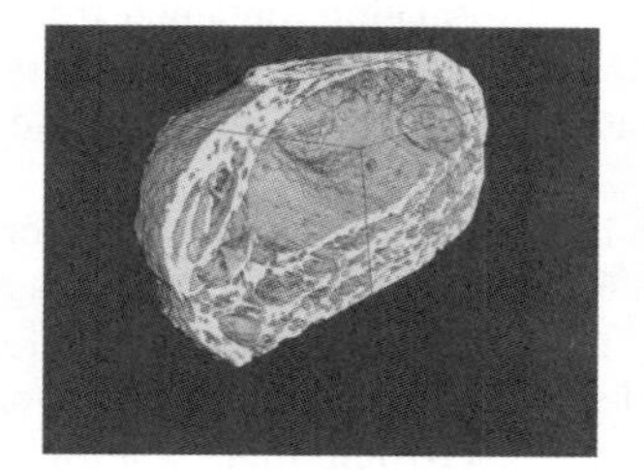

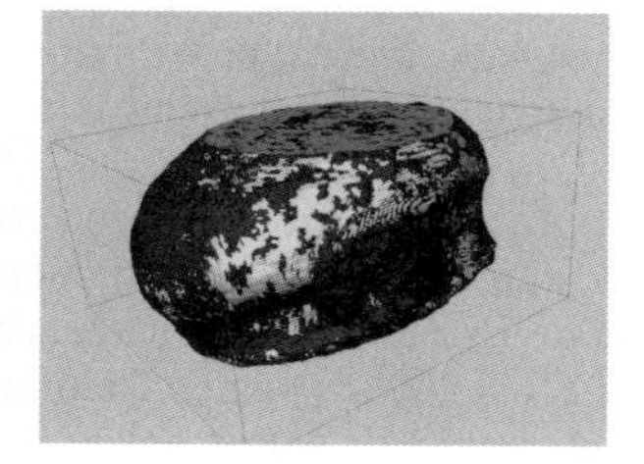

Figure 4 *3D XRT visualisation of internal structure (left) and non-uniform distribution of fat coating (dark grey - right).*

2.2 Assessment of sample hydration

Fig. 5 shows the experimental setup used to follow hydration of a soup inclusion once in contact with bulk water. To accomplish this work, the temporal resolution of a RARE measurement method was tuned to the rate of water redistribution at 25 °C and 60 °C. This was achieved by using short repetition time (TR = 200 ms) and RF of 4 and 8, which allow for temporal resolution of 3.5 min. and 2 min, respectively. The use of higher RF, which will allow for faster measurements was not convenient as it suppress signal from water saturating porous the matrix, while enhancing signal from bulk water surrounding it. A higher temperature was used for oil coated samples in order to properly assess temperature effects (fat melting). The obtained images with 3D spatial resolution of 0.3 mm allow for real-time visualisation of the spatial redistribution of moisture (Fig. 6, 7).

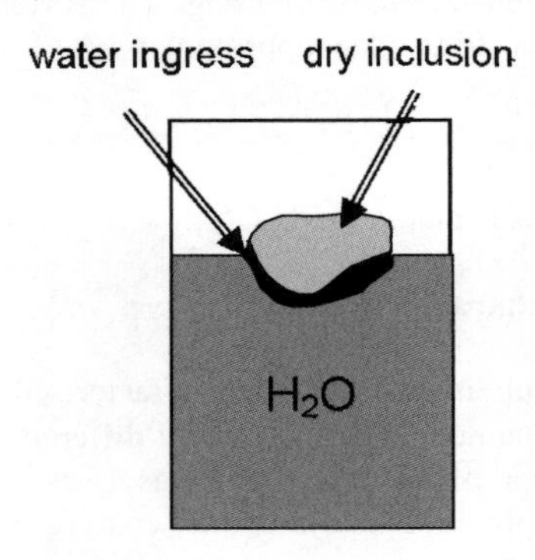

Figure 5 *Scheme of the MRI experiment on soup inclusion hydration.*

A first set of experiments was done on non-coated inclusions, in order to assess the effect of structure on hydration. Fig. 6 shows examples of fast, non-uniform hydration and slow, relatively homogenous hydration. Both samples were hydrated at the same conditions. Significant differences in hydration rate and behaviour could be attributed to differences in internal structure. In the first case, fast ingress of water occur which begins at specific region of a sample and was accompanied with swelling of the material. This at some point leads to "catastrophic failure" of the inclusion structure, and further fast hydration. In the second case, integrity of external part of sample retards hydration significantly. Correlation of the dynamic MR images with XRT structural information, obtained from dry samples prior to hydration experiments, prove that regions of a thin crust, in proximity of big pores were in most cases responsible for unwanted fast hydration (see Fig. 4 left and Fig. 6 top).

The second set of measurements was done on coated samples, at 60 °C, in order to assess effect of oil coating on retarding hydration. Fig. 7 shows set of MR images of coated inclusion. Due to its NMR properties (short T_1 as compared to water) and measurement setup (RF = 8, TR = 200 ms) signal from melted oil is enhanced over the bound and bulk water signal and appears on MR image as the brightest region. Images show predominantly slow homogenous ingress of melted oil into the soup inclusion, preventing hydration. However, it also shows a region on the surface from which faster hydration begins, leading to premature hydration of the entire sample. Correlation with XRT images proves that regions of the sample surface which were poorly coated were responsible for faster initial hydration (see Fig. 7).

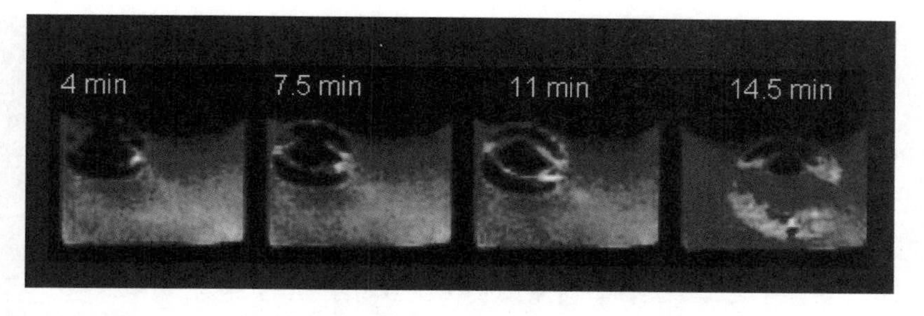

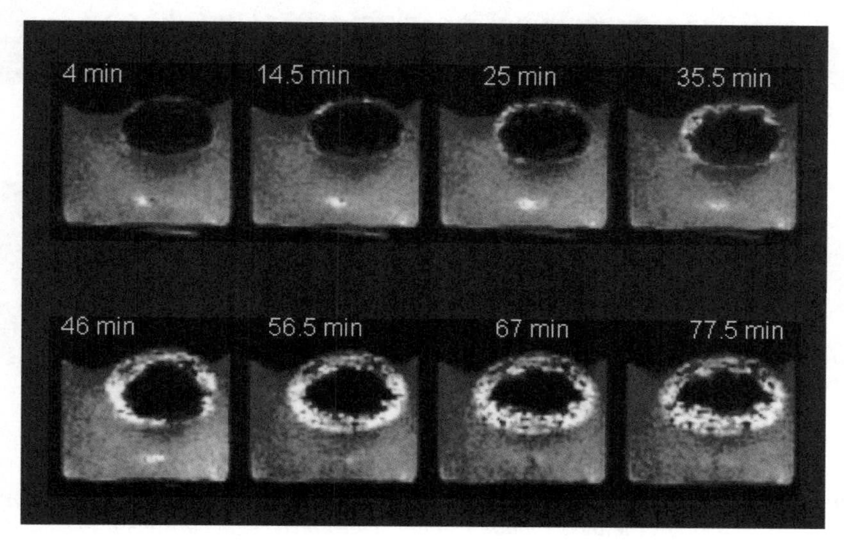

Figure 6

2D cross-sections from dynamic 3D MR images of soup inclusion hydration – fast, "catastrophic" water ingress (upper row) and slow homogenous diffusion of water into inclusion structure(middle and bottom rows).

4. CONCLUSION

The obtained results demonstrate the potential the combined MRI/XRT for monitoring the migration of moisture and accompanying structural changes within porous carbohydrate matrices. The superior spatial resolution of XRT allow for characterisation of sample structure, while MRI allow for dynamic visualisation of water ingress.

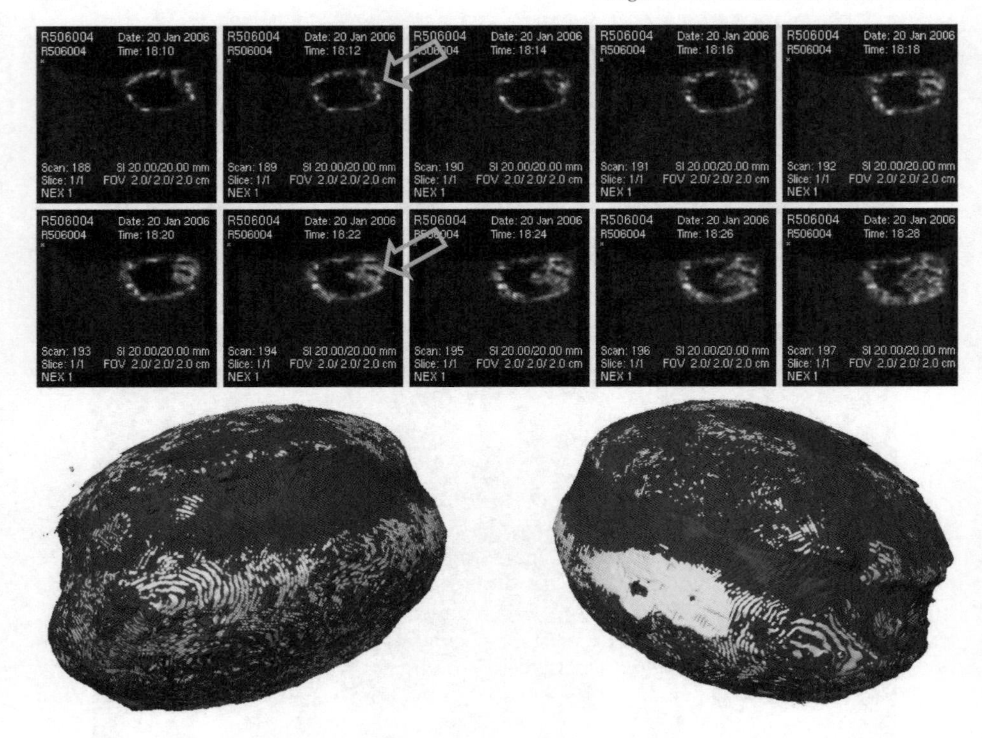

Figure 7 *Correlation of the dynamic MR images of soup inclusion hydration(top) with XRT images (bottom) showing uncoated region of sample surface (in light grey on XRT images) being reason for accelerated hydration.*

Acknowledgments:

Supported by European Community within the framework of a Marie Curie Intra-European Fellowship (MEIF-CT-2005-009475).

References

1 A.K. Horigane, S. Naito, M. Kurimoto, K. Irie, M. Yamada, H. Motoi, M. Yoshida, *Cereal. Chem.*, 2006, **83(3),** 235.
2 Mohorič, F. Vergeldt, E. Gerkema, A. de Jager, J. van Duynhoven, G. van Dalen, H. Van As, *J.Magn. Reson.*, 2004, **171(1),** 157.
3 G. van Dalen, H. Blonk, H. van Aalst, C.H. Luengo, *G.I.T. Imaging and Microscopy*, 2003, **3**, 18.

MRI STUDY OF POLENTA GELATINIZATION DURING COOKING

I. Serša, A. Sepe and U. Mikac

MRI Laboratory, Jožef Stefan Institute, Jamova 39, 1000 Ljubljana, Slovenia

1 INTRODUCTION

Polenta is a boiled, slow-cooked cornmeal "mush" typically made with ground dried cornmeal. Early forms of polenta originate from Roman times (pulmentum in Latin) when polenta was made of grain mush with faro (emmer wheat) and chestnut flour starches. Small amounts of these are still used today. Nowadays, polenta is traditional food in northern Italy from where many polenta recipes originate. Polenta is traditionally a slowly cooked dish, sometimes taking an hour or longer to cook. This has led to a profusion of shortcuts in cooking technique of which most rely on instant and precooked polenta. When boiled, polenta has smooth creamy textures, caused by the presence of starch molecules dissolved in the water. This process is also known as starch gelatinization.

Starch is an important constituent in many foods. It occurs in the leucoplasts of tubers, leaves, seeds and other portions of the plant in the form of granules that were formed during photosynthesis. Starch is composed of two polymers: amylose and amylopectin.[1,2] Amylopectin is the branched chained glucose polymer and contributes primarily to the viscosity of a prepared food, while amylose as a linear linked polymer has mainly gelling contribution. Amylose is approximately one-fourth the size of amylopectin. Amylose content in different starches ranges typically from 16 to 25% and the rest is amylopectin.

For every food starch there is a range of temperatures at which starch granules begin to lose their organized structure.[3] In corn starch - water dispersion during heating from 30°C to 50°C the granules preserve their appearance and are still held together by inter- and intra-molecular hydrogen bonding within and between the amylose and amylopectin. As the temperature rises to 50°C and 65°C hydrogen bonds are weakened and absorption of water within the granules is facilitated. This is accompanied also by a slight change in granule appearance. At 70°C the gelatinization process is the most extensive, while at 75°C to 80°C granules collapse and loose their shape.[4] During gelatinization starch granules first swell as they absorb water. When heated, the granules loose amylose that starts to fill intragranular space and some of granules collapse due to amylose loss.[5,6] When cooled, amylose molecules come closer and first junction zones are formed. Over time more and more amylose junction zones are formed that give rigidity to so formed starch gel.[7,8]

Process of gelatinization can be well followed by measuring viscosity changes of corn starch - water dispersion during heating.[9] In the present work we used NMR and MRI to follow the same process. Compared to other methods NMR offers large variety of different parameters that can be dynamically followed.[10] We employed MR microscopy, NMR relaxometry and diffusiometry to monitor starch gelatinization in polenta and tried to relate measured parameters with the degree of cooking.

2 MATERIALS AND METHODS

2.1 Sample Preparation

Two types of polenta samples were investigated: regular polenta and instant polenta both produced by the local food manufacturer (Droga, Slovenia). According to the manufacturer's recipe for regular polenta one unit of cornmeal should be mixed into four units of boiled water and polenta is done after 20 minutes cooking with occasional stirring, while instant polenta is done after just 2 minutes of identical procedure. Before every NMR/MRI measurement we prepared a set of polenta samples that were taken at various stages of cooking from a larger pot in which polenta was cooked according to the manufacturer's instructions. Regular polenta samples were acquired every 2 minutes from beginning (0 min) to the cooking end (20 min) while instant polenta samples were acquired every 30 seconds (from 0 min to 8 min). The samples were stored in small glass tubes with 8 mm diameter and 17 mm length and cooled down to room temperature before imaging or spectroscopy.

2.2 Hot Air Apparatus for Cooking Inside the NMR Magnet

In addition to sample preparation with quenching (stop cooking) procedure a simple home made apparatus was build that enables dynamical NMR studies while cooking samples in the magnet. The apparatus consisted of a pressurized air installation, nonmagnetic heater and a control unit with a temperature sensor, while NMR probes (Bruker, Germany) were already designed to withstand high temperatures and had inlet tubes for hot air and a temperature sensor. In operation pressurised air was led by hoses to a glass tube with the heater and from there to the RF probe with the temperature sensor. The control unit was regulating the heater power to maintain the temperature in the RF probe within a desired range which was usually 100°C ± 2°C. Samples were prepared in glass tubes (8 mm diameter and 17 mm length) that were filed to the top with cornmeal water mixture and sealed with stoppers having a small outlet hole for water vapour. The samples, at room temperature, were inserted in the RF probe horizontally and hot air heating was started simultaneously with dynamical NMR signal acquisition.

2.3 NMR/MRI Techniques

Experiments were performed on a system consisting of a 2.35 T horizontal bore superconducting magnet (Oxford Instruments, UK) equipped with accessories for MR microscopy (Bruker, Germany) with maximum gradients of 300 mT/m and NMR/MRI spectrometer (TecMag, USA).

Polenta at different stages of cooking was investigated by NMR relaxometry with Carr-Purcell-Meiboom-Gill (CPMG) multi echo train[11] for measuring T_2 relaxation times and with Inversion Recovery (IR) method for determining T_1 relaxation times. NMR diffusiometry, to measure Apparent Diffusion Coefficient (ADC) of the same samples by the Pulsed filed Gradients Spin Echo (PGSE) technique, was employed as well.[12] The techniques were used as spectroscopic methods, i.e., relaxation times and ADCs were measured from the whole sample, as well as in combination with 3D imaging (just for T_2 and ADC mapping).

Polenta gelatinization was studied also by MR microscopy with three different methods. High resolution proton density MR images were acquired using 3D spin-echo imaging technique with 10 ms echo time and 2000 ms repetition rate, at 2 signal averages and with imaging matrix 256 x 256 x 8 and filed of view 15 x 15 x 4 mm. Multi echo 3D imaging was performed to obtain images of T_2 relaxation time. The 3D images were acquired in 8 consecutive echoes with 14 ms inter-echo time, 1000 ms repetition time, with no averaging and at imaging matrix 256 x 256 x 32 and field of view 15 x 15 x 17 mm. The same sequence with just one echo and with (and without) readout gradient being constantly on between the excitation pulse and the echo was used as a PGSE sequence to obtain a diffusion weighted image with diffusion weighting of $b = 415$ s/mm^2 (and with $b = 0$ s/mm^2). Images of T_2 relaxation time were calculated in selected slices from the 3D multi echo image by ImageJ (NIH, USA) image processing software; the same software was used also to calculate ADC maps from diffusion weighted images of different b values. Dynamical 3D MR imaging was performed by the 3D RARE technique[13,14] with sequential k-space encoding in 64 echoes (each for one line in the first phase direction) of 1.6 ms inter-echo delay, 1000 ms repetition time, imaging matrix 64^3 and isotropic field of view of 15 mm. Second phase direction was sampled in separate scans with progressively increasing second phase gradient, so the total scan time was 76 s.

3 RESULTS AND DISSCUSSION

Effect of heat to cornmeal water mixture can be well seen in Figure 1 that depicts gelatinization process in regular polenta imaged dynamically by the 3D RARE technique. Initially (0 min) individual cornmeal grains are at the bottom of the 8 mm glass tube and water is above. In course of time hot water starts to migrate into the grains and transforms starch granules there. These swell and get coated with a layer of gelatinized starch that was released during disaggregation of the granules. As a result of the disaggregation the cornmeal grains swell as well and get coated by the gelatinized starch. This can be well seen in high resolution MR images in Figure 2 in regular polenta image at 30s of cooking and instant polenta image at 10 s of cooking. In these images, the grains are coated with starch gel that appears brighter than surrounding water due to shorter T_1 relaxation time, while the core of the grains appears dark as it contains practically no water. As cooking progresses the gel layer, gets thicker and the grains get smaller as more and more starch is transformed into gel surrounding the grains. Swelling of cornmeal grains can be seen in dynamic MR image in Figure 1 where after 4.5 minutes of cooking the grains are evenly distributed all over the sample tube. The effect of swelling can be seen in larger detail also in Figure 2 where the thickness of the gel layer around the grains increases while intact core of the grains gets smaller and smaller (regular polenta 30 s-4 min, instant polenta 10 s – 1 min). At the same time the amount of free water in between the grains decreases and completely disappears when starch granules completely disaggregate and polenta becomes a uniform gel. This, as evident in Figure 2, happens after 4 min of cooking for regular poneta and after 2 minutes for instant polenta.

Dynamical MR imaging of the sample while being cooked in Figure 1 has advantage in comparison to the high resolution experiment with sample quenching in Figure 2 for it enables sample imaging in an identical slice at different times, but unfortunately it has also several disadvantages. These are lower resolution in comparison to the standard 3D spin echo experiment and inability to follow the standard recipe. In our case cornmeal was mixed with cold water instead with boiling water and we were not able to stir polenta while cooking. This probably explains slower cooking in the dynamical experiment in Figure 1 in comparison to the quenching experiment in Figure 2. For example in Figure 1 individual cornmeal grains can be seen even after 13.5 minutes of cooking while in Figure 2 the grains are completely disaggregated after 4 minutes of cooking.

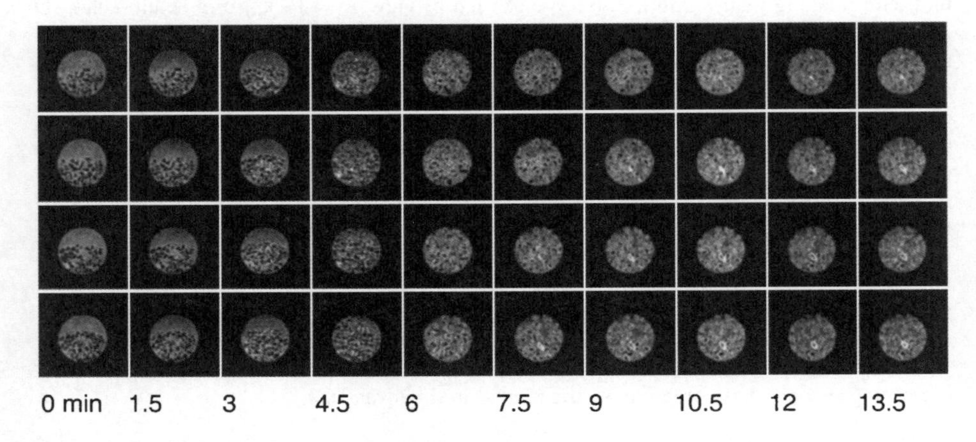

Figure 1 *Dynamical MR microscopy by the 3D RARE technique of polenta gelatinization in four consecutive 0.24 mm thick slices. The sample with regular polenta was cooked in 8 mm glass tube by hot air at 100 °C while being scanned.*

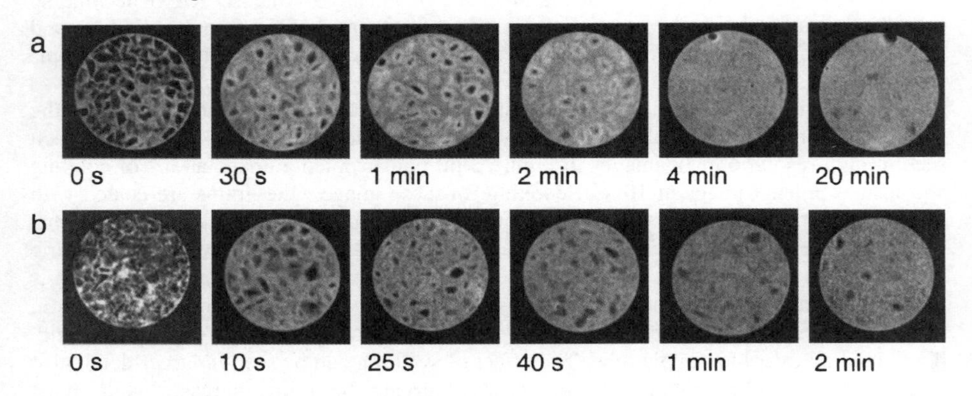

Figure 2 *High resolution MR images of regular (a) and instant (b) polenta at different cooking times. Samples were prepared in 8 mm glass tubes by the quenching procedure. The images were acquired by the 3D spin-echo technique at TE/TR=10/2000 ms and slice thickness 0.5 mm; single slices are shown.*

The polenta gelatinization process well reflects also in the change of NMR parameters: T_1, T_2, and the apparent diffusion coefficient (ADC). Figure 3 depicts how these parameters change with time in different stages of polenta gelatinization. The NMR parameters in Figure 3 were measured from the whole sample and therefore correspond to sort of sample average values. Nevertheless, they still give very interesting insight into progress of polenta cooking and may help determining how well and when polenta is done. T_1 relaxation time (Figure 3a) is at the beginning, when cornmeal grains did not start to disaggregate and are just mixed with cold water, quite low (just 900 ms), while T_2 relaxation time (Figure 3b) is at beginning lower than shortly after as well. This is most probably due to susceptibility effects of the grains to surrounding water. When gelatinization is the most intense, this is approximately after 1-2 minutes after beginning, both relaxation times, T_1 as well as T_2, reach the maximum value (1600 ms for T_1 and approximately 200 ms for T_2). At that time significant portion of cornmeal grains is already transformed into gel that coats the grains and therefore reduces susceptibility effect of the grains to free water between the grains. After 2 minutes gel network is starting to form. As the gel reduces water mobility it also changes relaxation times that decrease again and are at the end approximately equal to T_1 = 1300 ms and T_2 = 110 ms for both regular and instant polenta. Apparent diffusion coefficient (Figure 3c) also follows the trend of relaxation times. The coefficient is the highest at beginning (or immediately after for instant polenta) and then starts to decrease due to reduced water mobility caused by a forming gel. Denser the gel is lower is the water mobility and lower is ADC. At the end ADC is lower than $1.6 \cdot 10^{-9}$ m^2/s. Correlation graph in Figure 3d also proves that the T_2 relaxation time correlates well with the apparent diffusion coefficient.

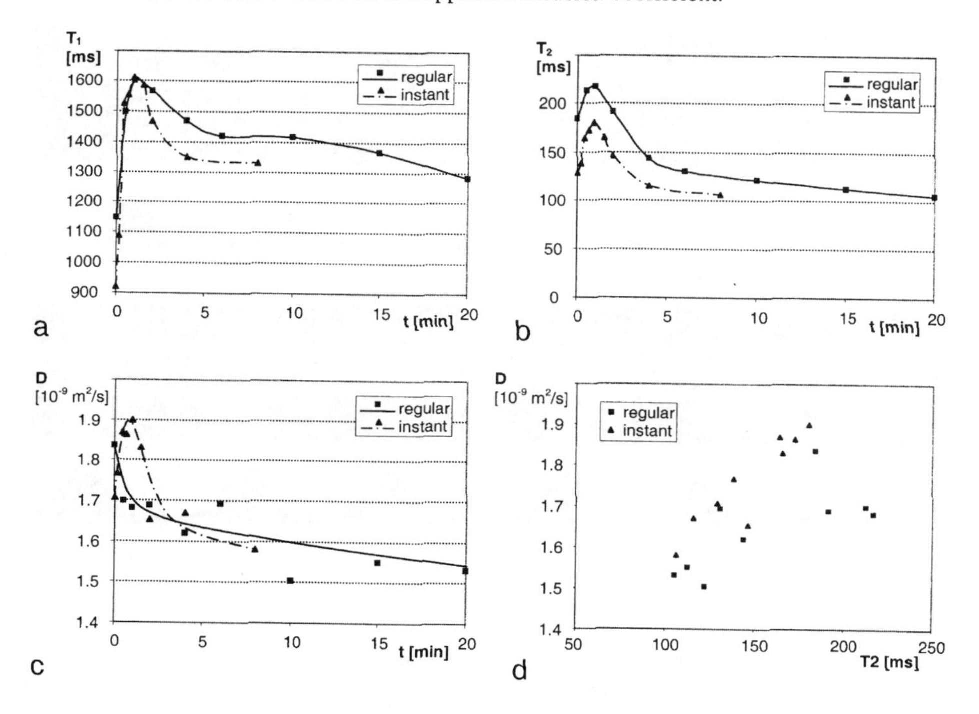

Figure 3 *Relaxation times T_1 (a), T_2 (b) and apparent diffusion coefficient (c) of regular (squares) and instant (triangles) polenta at different cooking times. Graph (d) depicts correlation between the apparent diffusion coefficient and the T_2 relaxation time.*

The disadvantage of measurements in Figure 3, i.e., that NMR parameters are measured from the whole sample, can be eliminated by measuring relaxation time and ADC maps of the sample. In these, a corresponding NMR parameter is measured for each image pixel. Figure 4 depicts images of the T_2 relaxation time and ADC at different cooking times of regular and instant polenta. T_2 relaxation times of polenta at cooking beginning (30 s for regular polenta and 10 s for instant polenta) vary a lot. The longest are that of free water (in some pixels T_2 is much longer than 300 ms) and is approximately 50 ms long within the grains. As cooking progresses the difference between T_2 relaxation time of the grains (if there are any) and the gel between them (formerly water) gets smaller. At the end (at 20 min for regular and 2 min for instant polenta) T_2 relaxation time distribution, which can be best seen in corresponding image histograms, has just one quite narrow peak at 100 ms which also well agrees with polenta becoming a uniform gel. Very similar is happening also with ADC but to a lower extent. At the beginning ADC of free water and that of the grains differs a lot ($2 \cdot 10^{-9}$ m^2/s vs. $1.1 \cdot 10^{-9}$ m^2/s), but as cooking progresses the differences get smaller which can be well seen in image histograms. For regular polenta, ADC image histogram (ADC distribution) is twice wider at the beginning than at the end. The difference for instant polenta is not as big.

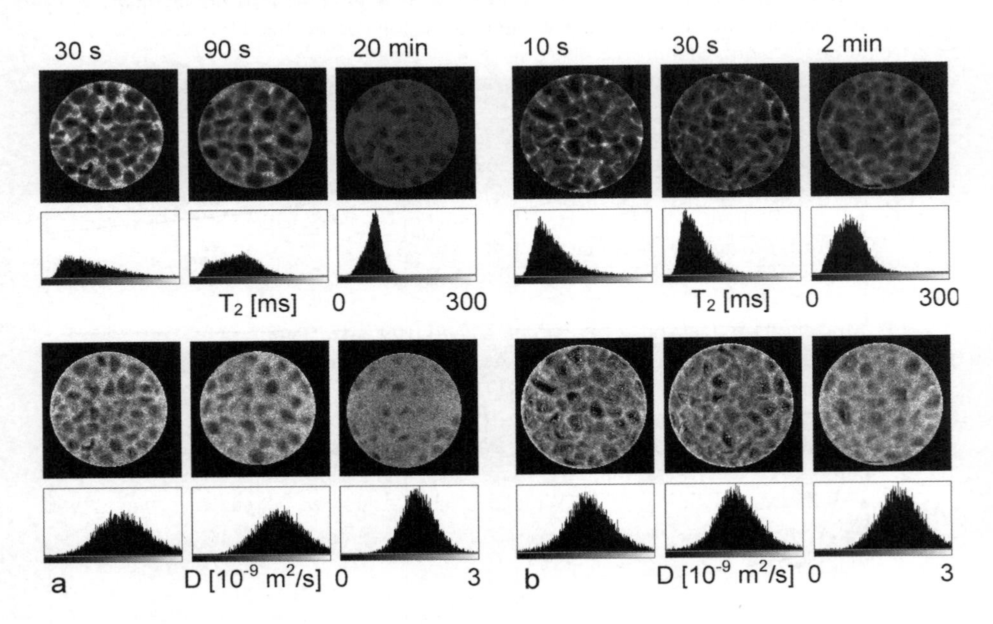

Figure 4 *T2 relaxation time images (upper row) and apparent diffusion coefficient images (lower row) of regular (a) and instant (b) polneta at different cooking times. Distribution of T_2 relaxation times or apparent diffusion coefficients across the image is depicted by the image histogram.*

Relaxation time and diffusion measurements indicate that when polenta (regular as well as instant) is done T_1 drops to approximately 1300 ms, T_2 to 110 ms and ADC drops below $1.6 \cdot 10^{-9}$ m^2/s. The followed NMR parameters also indicate that instant polenta is done approximately four times faster than regular polenta.

4 CONCLUSIONS

NMR parameters such as relaxation times T_1, T_2 and the apparent diffusion coefficient are convenient for quantitative analysis of a food preparation process. In combination with high resolution MR imaging they may give very useful insight into changes affecting grained or powdered food during the preparation process. In our study with polenta we have demonstrated that coking progress can be well monitored by these parameters and can also help determining how well and when the food is done.

Acknowledgements

The authors thank Dr. Aleš Mohorič, University of Ljubljana for many stimulating discussions and constructive suggestions.

References

1. D.J. Gallant, B. Bouchet and P. M. Baldwin, *Carbohydr. Polym.*, 1997, **32**, 177.
2 P.J. Frazier, A.M. Donald and P. Richmond, Starch: Structure and Functionality, Royal Soc. Chem., Cambridge, 1997.
3 N. Singh, J. Singh, L. Kaur, N.S. Sodhi, and B. S. Gill, *Food Chemistry*, 2003, **81**, 219.
4 W.S. Ratnayake and D.S. Jackson, *J. Agric. Food Chem.*, 2006, **54**, 3712.
5 J.W. Donovan, *Biopolymers*, 1979, **18**, 263.
6 R.F. Tester and W.R. Morrison, *Cereal Chemistry*, 1990, **67**, 558.
7 S.G. ring, P. Collona, K.J. I'Anson, M.T. Kalichevsky, M.J. Miles, V.J. Morris and P.D. Orford, *Carbohydrate Research*, 1987, **162**, 277.
8 M.J. Miles, V.J. Morris, P.D. Orford and S.G. Ring, *Carbohydrate Research*, 1985, **135**, 271.
9 J.P. Mua and D.S. Jackson, *J. Agr. Food Chem.*, 1997, **45**, 3848.
10 A. Mohoric, F. Vergeldt, E. Gerkema, A. de Jager, J. van Duynhoven, G. van Dalen, and H. Van As, *J. Magn. Reson.*, 2004, **171**, 157.
11 S. Meiboom and D. Gill, *Rev. Sci. Instr.*, 1959, **29**, 688.
12 E.O. Stejskal and J.E. Tanner, *J. Chem. Phys.*, 1965, **42**, 288.
13 J. Henning, A. Nauert, H. Freidburg, *Magn. Res. Med.*, 1986, **3**, 823.
14 R.V. Mulkerrn, S.T.S. Wong, C. Winalski, and F.A. Jolesz, *Magn. Res. Imag.*, 1990, **8**, 557.

THE MELTING BEHAVIOUR OF LARD IN "DANISH STYLE" LIVER PÂTÉ AS MEASURED BY DSC AND TD-NMR

G. Svenstrup[1], E. Micklander[1,2], J. Risbo[1] and I.A. Farhat[3]

[1] Department of Food Science, Centre for Advanced Food Studies, Royal Veterinary and Agricultural University, DK-1958 Frederiksberg, Denmark.
[2] Current affiliation: AgResearch Ltd, Ruakura Research Centre, East Street, Private Bag 3123, Hamilton, New Zealand
[3] Division of Food Sciences, School of Biosciences, University of Nottingham, Sutton Bonington Campus, Loughborough, LE12 5RD, UK

1 INTRODUCTION

The presence of fat in foods influences their physical properties (processing behaviour, texture and mouth feel, flavour release, etc.). Fat crystallization plays a major role in the processing of chocolate (1-3), margarine and shortenings (4-6). While many reports relating to such systems can be found (1-6), much less is known regarding the physical and chemical properties of fats in cooked meat products. Knowledge that would be useful for the optimization of organoleptic quality of various cooked meat products.

The temperature at which a fat system melts or crystallises depends on:

- Chemical structure, i.e. the aliphatic chain length, its degree of saturation and the extent of acylation (mono-, di- or triacylglycerol).
- Crystal packing. The crystal crystalline state of fats is polymorphic, i.e. the same fat could crystallise under different crystalline forms (polymorphs). The 3 main polymorphic forms commonly encountered in most vegetable and animal fats are, in order of increasing thermal stability: α, β', and β. Their melting point increases with the increase of their crystal packing in the order: α, β', β.

Crystallization in simple emulsions (oil in water) has been investigated in details (7). However, for meat products, like pâté the emulsion is more complex as it contains many different components such as proteins (principally from the liver), carbohydrates (largely from the added flour), etc. and exhibits heterogeneities over a range of distance scales.

Differential Scanning Calorimetry (DSC) is often used as the method of choice to describe the melting behaviour of edible fats, such as beef fat and margarine (8), chocolate (1), and pork fat (9).

Proton time domain nuclear magnetic resonance (TD-NMR) often implemented on low magnetic field spectrometers (typically ~0.5 T) proved to be a valuable tool for the determination of solid fat content and total fat content in dry foods. NMR is essentially sensitive to proton molecular dynamics. In meat systems, the technique -F has been used to measure "juiciness" and was successfully applied to measure the oil and water contents in fish (10) , determine fat content of meat (11) and to predict water holding capacity of pork (12, 13).

In this study, we have investigated the crystallization/melting behaviour of pure lard *vs.* lard in an emulsion – i.e. in a Danish style liver pâté using DSC and TD-NMR.

2 MATERIALS AND METHODS

Lard was obtained from a commercial production of slaughter pigs of DLY cross bread - using Duroc as the terminal boar and a Danish Landrace Yorkshire as the female pig. The physical properties of the fat were investigated in pure minced lard and in lard as an ingredient of liver pâté.

The recipe for the liver pâté can be found in Table 1. The lard was boiled in water for 30 minutes. The liver was minced with the dried onion. The other ingredients (wheat flour, dried milk, salt, glucose, white pepper, and thyme) were minced separately, and the boiling liver preparation was gradually added while stirring at high speed. The minced liver was thereafter added during low speed stirring. The liver pâté samples were kept uncooked and were stored at -20 °C until analysis along with the samples of pure lard.

Table 1 *Liver pâté, list of ingredients.*

Ingredients	Quantity	Ingredients	Quantity
Lard	37.0 %	Salt	0.93 %
Liver	30.4 %	Dried onions	0.56 %
Water	25.2 %	Glucose	0.19 %
Wheat flour	3.1 %	White pepper	0.15 %
Dried milk	2.5 %	Thyme	0.03 %

The thermal properties were measured using a differential scanning calorimeter DSC 820, Mettler Toledo (Schwerzenbach, Switzerland) based on the heat flux principle and cooled with liquid nitrogen. The heat flow and temperature were calibrated with indium (Tm = 156.6 °C, ΔH_{fus} = 28.5 J/g) and zinc (T_m = 419.5 °C, ΔH_{fus} = 107.5 J/g) as standards. The low temperature calibration was checked with decane (T_m = -29.66 °C) and cyclohexane (T_m = 6.47 °C). The sample was hermetically sealed in an aluminium sample pan of 40 µl. An empty, hermetically sealed aluminium pan identical to the sample pan was used as reference. Experimental conditions were identical for all products: The samples were heated to 80 °C and held at this temperature for 20 minutes to ensure that the fat was completely melted and all the nuclei fully eliminated (14). The samples were subsequently cooled to 5 °C at a cooling rate = 0.5 K/min. The melting thermograms were recorded during the heating of the samples to 65 °C with heating rates of 1.0 °C/min.

The TD-NMR measurements were performed using a MARAN Benchtop pulsed NMR spectrometer (Oxford Instruments, UK) operating at 23 MHz and equipped with an 18 mm variable temperature probe. Transverse relaxation times, T_2, were determined using the Carr-Purcell-Meiboom-Gill sequence (CPMG). The sample temperature in the probe-head was controlled by a continuous flow of dried air. The dead time of the instrument was 9 µs. Approximately 5 g of sample was sealed in 15 mm glass tube, which was inserted inside a standard 18 mm NMR tube located in the spectrometer probe-head. This was done to minimise headspace and avoid any changes in water content during the heating/cooling cycles. The CPMG parameters were: relaxation delay (RD) = 2 s, point per echo (PECH) = 1, tau = 250 µs, numbers of echos (NECH) = 8192. The experimental conditions were identical for all samples: every measurement at each temperature took 12 minutes, of these, 5 minutes was the time allowed to ensure thermal equilibrium across the sample. The samples were heated from 25 °C to 80 °C. 2 measurements were executed at 80 °C (to ensure the totally melting of the fat). The samples were subsequently cooled to 5 °C

(Cooling rate ~ 0.4 K/min, comparable to the cooling rate of 0.5 K/min used for the DSC). The samples were subsequently reheated to 59 °C and measurements were carried out for every 2 K.

MATLAB version 6.5 (release 14, The MathWorks, Inc., Natick, MA, USA) was used for data analysis. The relaxation spectra were obtained by fitting exponential distributions to the CPMG decays:

$$M(t) = \sum_{i=1}^{N} M_i \cdot \exp\left(\frac{-t}{T_{2,i}}\right) \qquad (1)$$

where $M(t)$ is the residual magnetization at time t described as a sum of N mono-exponentials each with the initial signal magnitude M_i and the relaxation time $T_{2,i}$. In distributed exponential analysis, a distribution of magnitudes M_i is obtained for a large number of $T_{2,i}$ (Equation 1). In this work 256 $T_{2,i}$ values sampled logarithmically in the time constant interval 10^{-1}-10^4 ms were used.

3 RESULTS AND DISCUSION

3.1 Comparing the thermal transitions of lard and liver pâté

After cooking and cooling the pure lard and liver pâté in the DSC, the samples were reheated from 5°C to 65°C in order to compare the melting behaviour of the pure pork fat, the lard, to that of the fat in the liver pâté. The endotherms (Figure 1) showed similar overall patterns with the main thermal transition being a melting endotherm centred at ~30°C for the lard and at a slightly higher temperature (~32°C) for the liver pâté.

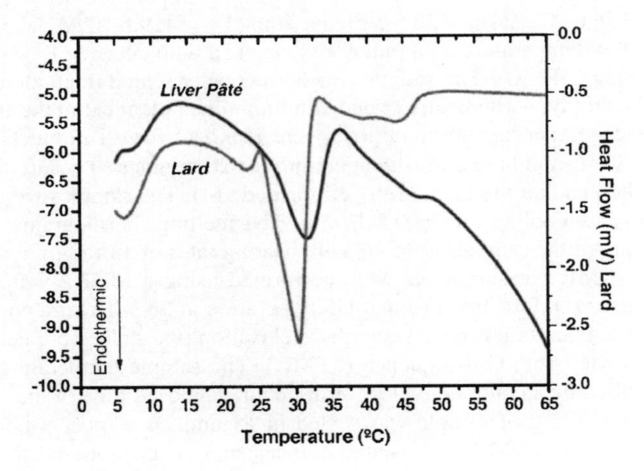

Figure 1 *DSC thermograms of pure lard and liver pâté made of lard obtained by reheating at 1.0 K/ min.*

The 3 main differences between the thermal responses of the 2 materials were:

- The size of the melting endotherm: this is expected as the liver pâté contained only 37% pork fat

- The melting temperature range for the main endotherm, which was broader for the pâté (higher peak and end temperatures).
- The minor melting events are slightly more resolved in the lard. This could be due to the more complex nature of the pâté both in terms of composition and microstructure.

It is interesting to note that while the peak and end temperature of the pâté melting were higher than those for the lard, there is little evidence of difference in the onset of melting, which is usually regarded as the "true" melting point. The spread in the melting range could therefore be due to the microstructure of the pâté, which could cause thermal lags, and spread of the melting range, than to polymorphic differences of the crystalline fat.

3.2 TD-NMR study of the melting behaviour of pure lard on reheating

Contour plots of the evolution of the relaxation spectra of pure lard obtained through the distributed exponential fitting of the TD-NMR CPMG decays recorded during reheating in the spectrometer probe-head after a first heating cycle are shown in Figure 2. A weak CPMG signal is observed at low temperatures reflecting the dominance of rigid behaviour at such temperatures where most of the NMR signal of the Free Induction Decay (FID) would have decayed before the refocusing 180° pulse is applied. As the temperature increases, the overall CPMG signal increased slightly and when the temperature exceeded 20°C, a relaxation distribution centred at ~28 ms became apparent. When the temperature increased above 30°C, a second relaxation distribution (T_2~120oC) was clearly observed. The magnitudes and relaxation times of these two proton populations, labelled P2 and P3 increased with temperatures and the T_2s reached values of 90 ms and 300 ms, respectively, at the highest experimental temperature of 59 °C.

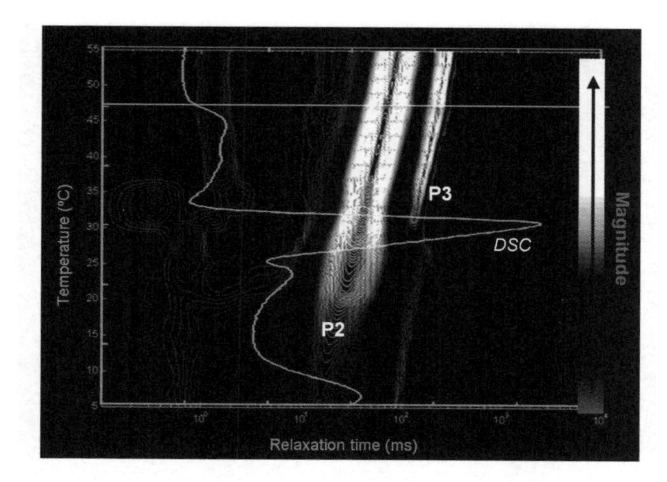

Figure 2 *Contour plots of distributed exponential fitting of CPMG data acquired during the reheating of pure lard. The equivalent DSC thermogram is superimposed for comparison (rotated by 90°). The 2 relaxing proton populations are labeled as P2 and P3.*

The increases in signal above 20°C and 30°C reflects the appearance of liquid fat and is in agreement with the DSC result which showed a melting endotherm with an onset temperature of ~20°C and a peak temperature at ~23°C and another more pronounced one with an onset temperature ~25°C and a peak temperature at ~30°C.

These 2 melting events could reflect the presence of 2 polymorphs or different chemical structures or both. Is usually difficult to deconvolute the contributions of these 2 factors without detailed physical (x-ray diffraction) and chemical (composition) analyses.

It is tempting to assign these 2 endotherms to P2 and P3 respectively, however a fully quantitative comparison between DSC and TD-NMR should take into account:

- The dependence of melting enthalpy on the polymorphic form. High Tm polymorphs often exhibit a higher melting enthalpy (15).
- The need to account for proton densities. This is difficult due to the limited chemical information regarding the nature of the components contributing to each population. This effect is expected to be minor, as changes in the degree of saturation of fatty acids would have a limited impact on their proton densities.

Assuming that the comparison of DSC and TD-NMR data is a valid one, the results indicate that at the same temperature (in the melt), TD-NMR is able to resolve the mobilities of liquid fats originating from solids of different melting temperatures and thus of different degrees of saturation. This observation merits further investigation.

3.2 Comparing the reheating melting behaviour of lard and liver pâté by TD-NMR

The comparison of the TD-NMR relaxation behaviour on reheating liver pâté (Figure 3) to that of the pure lard (Figure 2) clearly reveals that:

- The P2 and P3 population are present in both systems despite a weaker contribution in the pâté, most likely do to the fact that the pâté contained only 37% lard.
- The existence of a pronounced proton population centred at a T_2~8ms.

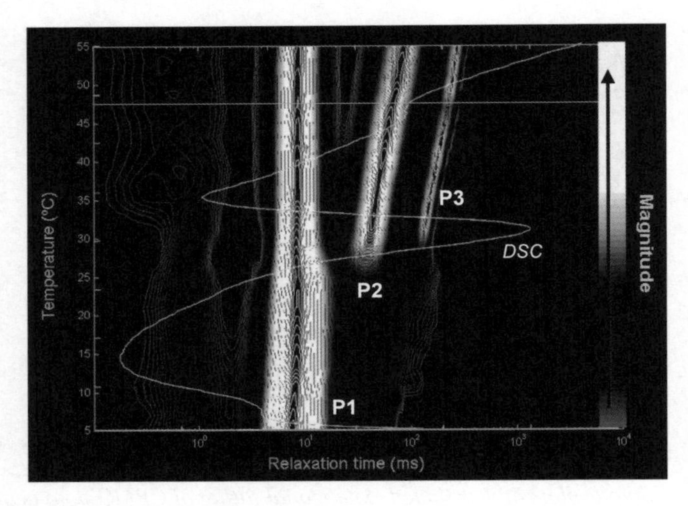

Figure 3 *Contour plots of distributed exponential fitting of CPMG data acquired during the reheating of liver pâté. The equivalent DSC thermograms are superimposed for comparison (rotated by 90°). Three relaxing proton populations P1, P2 and P3 are indicated.*

The behaviour and origin of P2 and P3 was discussed above for the pure lard and it is proposed that these are the same populations and resulting NMR signals in the case of the liver pâté. A slight difference in the case of the liver pâté is that P2 and P3 seem to become noticeable around the same temperature range, i.e. just under 30°C. This is compatible with DSC results, which did not show a clear resolution of 2 endotherms as was the case for the pure lard.
The main difference between the pure lard and the pâté from an NMR relaxometry viewpoint, and specifically in terms of the CPMG experiment which probes liquid-like behaviour, is the presence of water (25.2% added in addition to that contained in the other ingredient, especially the liver). The presence of a significant amount of hydrophilic solutes such as proteins and carbohydrates increase the relaxation rate of water by reducing its mobility and the possibility of proton exchange. Relaxation times of water in highly concentrated foods typically range from few milliseconds to tens of milliseconds (see for example Chapter by Sereno et al.). The relaxation of P1 showed little dependence on temperature while its overall magnitude decreased with increasing temperature, quite significantly around 30°C, i.e. in the temperature range where P2 and P3 became detectable. The implication of this is that with contributions from several mobile components to the relaxation spectrum, the fitting algorithm may have artificially overlapped P1 and P2 signals, i.e. a fraction if P2 may be a residual contribution from P1.

4 CONLUSION

The results show the benefit of complementing an established method such as DSC with TD-NMR when studying complex systems. DSC, which is often considered as the technique of choice to study phase transitions such as fat melting, it provides no information on the components, which do not undergo a thermodynamic transformation, as is the case for water in the pâté over the temperature range of interest (5°C-60°C) or the melted liquid fats (above their individual melting temperatures). TD-NMR experiments where the samples were cooked and reheated *in-situ* provided such information by clearly separating two liquid fat components resulting from the melting of 2 different fat fractions in the lard in its pure state or within the liver pâté. This observation suggests that the multiple melting behaviour observed by DSC reflects a "fractionation" of fat components with different chemical structures rather than the melting of similar structure in different polymorphic forms. The signal from water in the pâté was resolved from that of the fat (in particular at low temperatures) suggesting that a quantitative analysis of fat and water in such complex system is possible but would need to rely on Free Induction Decay data (total signal) and CPMG at low temperature (above freezing of course) for the water signal, or acquiring CPMG data at 2 temperatures, 1 below the onset of fat melting and beyond the completion of melting.

References

1. S.Bolliger, B.Breitschuh, M.Stranzinger, T.Wagner, and E.J.Windhab. Journal of Food Engineering **35,** 281 (1998).
2. N.Brunello, S.E.McGauley, and A.G.Marangoni. Lebensmittel-Wissenschaft und Technologie - Food Science and Technology **36,** 525 (2003).
3. A.G.Marangoni and S.E.McGauley. Crystal Growth & Design **3,** 95 (2003).

4. B.S.Ghotra, S.D.Dyal, and S.S.Narine. Food Research International **35,** 1015 (2002).
5. C.Moziar, J.M.deMan, and L.deMan. Journal of Canadian Institute of Food Science and Technology **22,** 238 (1989).
6. S.S.Narine and K.L.Humphrey. Food Research International **37,** 28 (2004).
7. S.A.Vanapalli, J.Palanuwech, and J.N.Coupland. Colloids and Surfaces A: Physicochemical and Engineering Aspects **204,** 227 (2002).
8. N.Aktas and M.Kaya. Journal of Thermal Analysis and Calorimetry **66,** 795 (2001).
9. G.Svenstrup, D.Bruggemann, L.Kristensen, J.Risbo, and L.H.Skibsted. European Journal of Lipid Science and Technology **107,** 607 (2005).
10. S.M.Jepsen, H.T.Pedersen, and S.B.Engelsen. Journal of the Science of Food and Agriculture **79,** 1793 (1999).
11. H.T.Pedersen, H.Berg, F.Lundby, and S.B.Engelsen. Innovative Food Science & Emerging Technologies **2,** 87 (2001).
12. H.C.Bertram, H.J.Andersen, and A.Karlsson. Meat Science **57,** 125 (2001).
13. H.C.Bertram, A.Schäfer, K.Rosenvold, and H.J.Andersen. Meat Science **66,** 915 (2004).
14. J.W.Litwinenko, A.M.Rojas, L.N.Gerschenson, and A.G.Marangoni. Journal of the American Oil Chemists' Society **79,** 647 (2002).

New Techniques and Novel Data Analysis and Exploitation

MOTIONAL RELATIVITY AND NOVEL NMR SENSORS

B. Hills, K. Wright, N. Marigheto and D. Hibberd

Institute of Food Research, Norwich Research Park, Colney, Norwich NR4 7UA, UK

1 INTRODUCTION

Conventional NMR (and MRI) is usually performed on stationary samples using rapidly switched radiofrequency (RF) and magnetic field gradients to excite and modulate an NMR response. However there are many circumstances where there are advantages in performing NMR in alternative ways made possible by the principle of "motional relativity". It is convenient to discuss translational and rotational motion separately as they gives rise to quite distinct methodologies. Translational relativity, to be discussed in section 2, is particularly appropriate for discrete samples travelling at uniform velocity which is a situation often found on industrial conveyors. In contrast, rotational relativity is particularly appropriate in Halbach magnet arrays and this will be the focus of section 3.

2 TRANSLATIONAL RELATIVITY AND ON-LINE NMR SENSORS

The principles underpinning translational relativity have been described in several recent publications[1-3] so need not be repeated here. Translational relativity allows the time dependence to be taken off the pulsed radiofrequency fields and switched gradients by exploiting the time dependence of the sample motion itself. In other words, an RF excitation can be achieved by translating the sample through a finite region of constant (non-switched) RF radiation (a process called the Sample Translation or "ST" mode in reference 1). Likewise one- (and potentially two-) dimensional imaging can be performed by translating the sample through regions of constant (i.e. non-switched) magnetic field gradient, using either the ST mode or RF pulse excitation. As discussed in reference 1, the ST mode is only really practical with the simplest NMR protocols, such as an FID or Hahn echo. Sequences involving long trains of RF excitations are best performed in the pulsed mode. Whichever excitation and modulation mode is used it is necessary to replace the conventional phase-cycled, multiple repetition pulse sequences with equivalent fast, single-shot sequences because each sample only travels once through the magnet. However, devising robust, ultrafast single-shot on-line pulse sequences, especially for 2-dimensional relaxometry and imaging, is far from trivial and will therefore be the focus of the rest of this section. Other aspects, such as hardware design and the need to pre-polarise samples with long T_1s have been discussed previously[3].

2.1 Ultra-fast, Single-shot On-line Pulse Sequences

In the absence of phase cycling the Hahn-echo or CPMG spin-echo train is inherently single-shot and provides ready on-line measurement of transverse relaxation times. However sample speed as well as the complications of "translational dephasing"[1] and RF and B_0 field inhomogeneity[1] set an upper limit on the measurable T_2. The single-shot FIRE sequence[1] has been successfully used to measure both T_1 and short T_2's although the data processing is cumbersome. These sequences use hard, non-selective RF pulses but the recent development of ultrafast multidimensional high resolution NMR[4] based on slice-selection with soft RF pulses suggests that a similar approach might, in principle, be valuable in developing new single-shot sequences appropriate for the on-line situation. Figure 1 illustrates the idea. A conventional 1-dimensional T_1 (or 2-dimensional T_1-T_2) measurement is very slow because of the need to repeatedly step out the inversion recovery time, t_1, which, to avoid saturation, necessitates a delay of $5T_1$ between each repetition. However, by using slice-selective RF 180^0 pulses in a fixed gradient, different inversion recovery times can be associated with different slices in the sample and so can be acquired in a single-shot. Each slice has a different t_1 because the slice selection pulse is progressively delayed. The signal in each slice can be separately resolved by acquiring an FID (or Hahn echo with short echo time) in a constant G_z gradient. The points in the one-dimensional profile obtained by Fourier transformation are then associated with different slices and therefore different inversion recovery times, allowing T_1 to be extracted directly from the profile. If instead of a simple FID or Hahn echo, a complete CPMG echo train is acquired in the read-out gradient, then this provides a second "t_2" dimension associated with T_2 relaxation and the sequence allows the acquisition of a single-shot, ultrafast 2-dimensional T_1-T_2 spectrum. In the on-line situation the slice selection would be performed with a rapid series of selective RF pulses as the sample travels through a non-switched linear gradient oriented along the direction of sample motion.

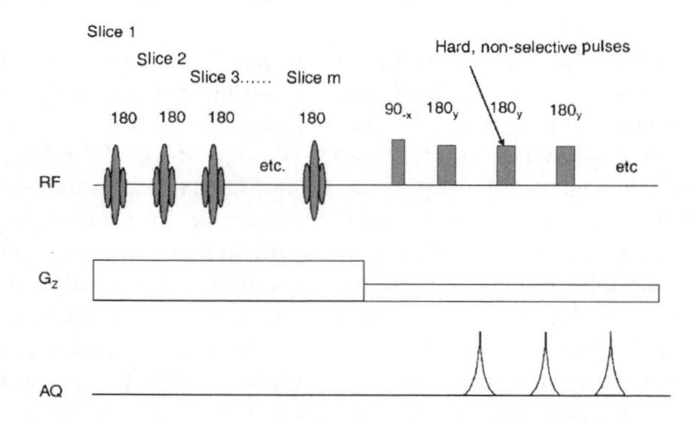

Figure 1 *The proposed single-shot, ultrafast T_1-T_2 pulse sequence.*

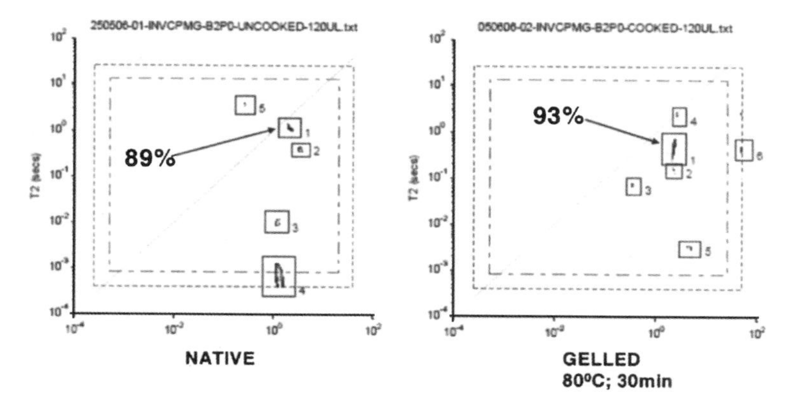

Figure 2 *The T_1-T_2 spectra of native and gelled 2%w/w aqueous BSA, pH 6.8 in 0.11 molar NaCl and acquired at 100MHz. Note the high intensity of the water peak 1.*

Although this single-shot, ultrafast on-line (and off-line) T_1-T_2 pulse sequence appears feasible there is another problem that arises in water-rich samples. Namely the high relative intensity of the water peak tends to obscure the more interesting peaks arising from dissolved solutes and biopolymers. Figure 2, acquired at 100MHz from a 2% BSA solution and gel, illustrates the problem. The water peak (peak 1) is seen to comprise 89 and 93% of the total integrated peak intensity respectively. Clearly it would be desirable to incorporate some form of water-suppression into the single-shot T_1-T_2 sequence. A possible water-suppression protocol which exploits the high diffusivity of water molecules relative to dissolved solutes, is illustrated in figure 3. It comprises three parts, namely an initial water-suppression PGSE sequence, followed by a modified pulsed gradient stimulated echo sequence and finally a CPMG sequence. The gradients in the initial PGSE part of the sequence are first adjusted to suppress the signal from the more mobile water.

The stimulated echo part of the sequence in figure 3 not only provides the variable t_1-dimension but also ensures that any fresh water magnetisation created by longitudinal relaxation during the t_1 delay is spoiled by the second gradient pulse of the stimulated echo sub-sequence. The same "slice-selection" concept can be used to develop single-shot, versions of other multidimensional relaxation/diffusion experiments such as the T_2-D sequence. Water suppression by diffusion is also the key to the single-shot, on-line measurement of the oil and soluble solid (Brix) contents of intact fruit. Figure 4 shows the pulse sequence where the first echo 'a' is used as a measure of the sample size and, because of water suppression, the second echo 'b' is dominated by the more slowly diffusing molecules, so that the ratio (b/a) is proportional to the oil or solute content. A weak constant gradient is added to improve signal/noise by sharpening the echoes. The experimental linear calibration curves for the on-line determination of oil content in intact avocado tissue[5] and of the Brix of apple and strawberry[6] are shown in figure 5.

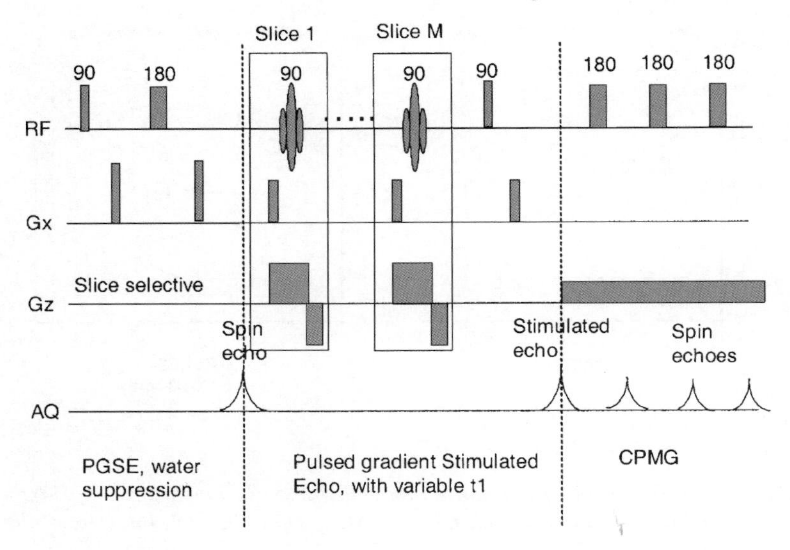

Figure 3 *The proposed, single-shot, ultrafast water-suppressed T_1-T_2 pulse sequence*

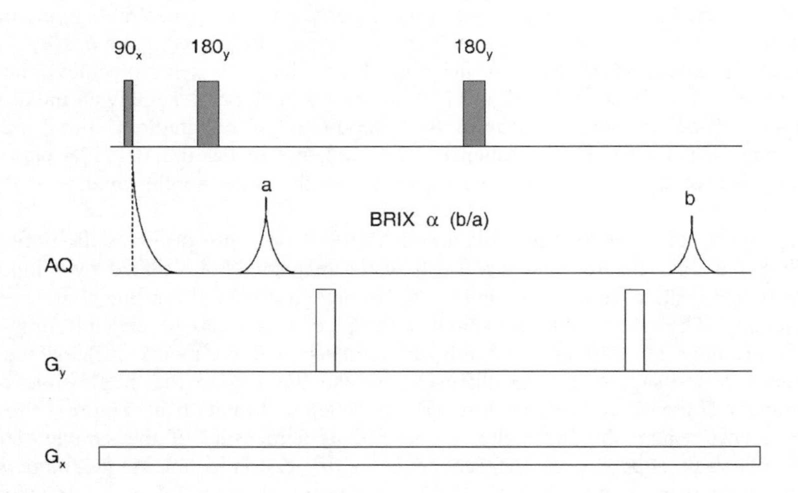

Figure 4 *The single-shot, on-line, Brix pulse sequence*

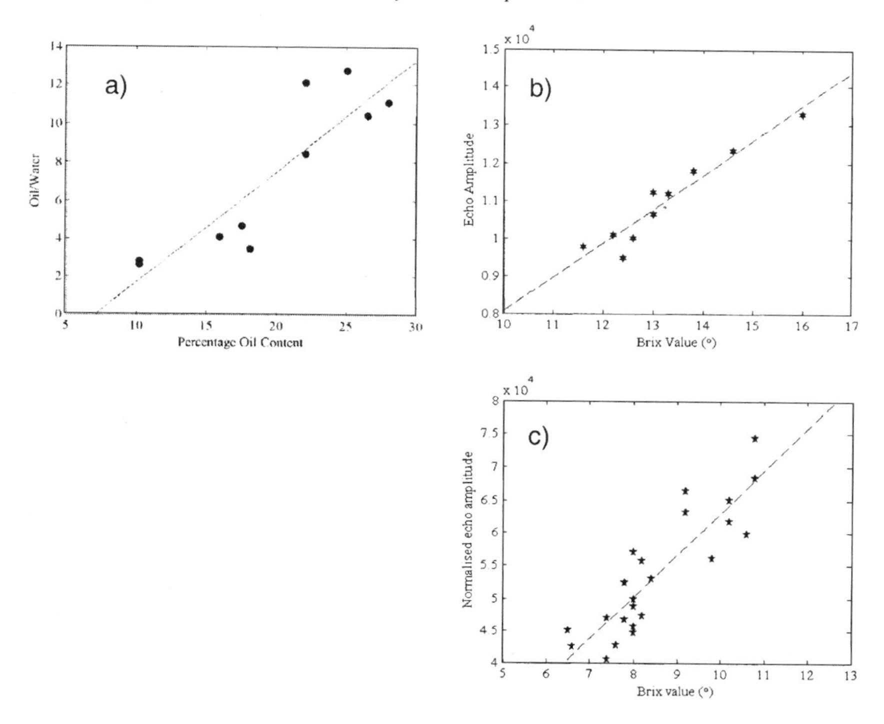

Figure 5 *The single-shot calibration curves for a) avocado oil content, (from ref. 5) b) Brix in apples and c) Brix in strawberries (from ref. 6).*

Single-shot, on-line, one-dimensional image profiles are, of course, straightforward to acquire because they merely require the Fourier transformation of a spin echo acquired in a linear gradient[1]. However, when the sample is moving at high velocity (>1m/s) and the gradient is oriented along the direction of motion a motional correction must be made[1]. Extending the single-shot protocol to 2 or 3-dimensional on-line imaging is more challenging but may be possible with an on-line version of the COMSPIRA pulse sequence[7]. COMSPIRA maps k-space in a single-shot Archimedean screw trajectory by simultaneously applying two sinusoidally oscillating transverse gradients, G_x and G_y whose amplitude increases linearly and between which there is a constant phase difference (see figure 6).

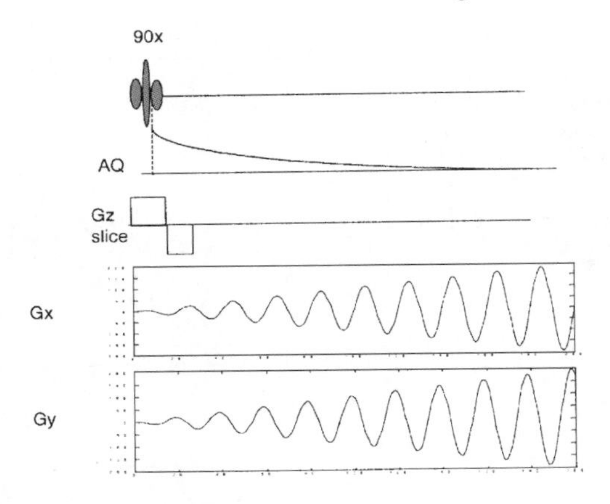

Figure 6 *The ultrafast COMSPIRA pulse sequence.*

Translational relativity means that the same Archimedean trajectory could be achieved using a single extended constant (non-oscillating, non-switched) transverse gradient but twisted in such a way that the sample experiences a changing gradient direction as it moves through the coil. Figure 7 shows a schematic of such a coil and simulations confirm the rotating direction of the transverse gradient (figure 7 insert). In the single-shot, on-line situation the 2D profile would be limited to a single slice through the sample obtained in the usual way by a selective RF pulse in a constant G_z gradient although the use of several RF coils and/or spin or gradient echo methods might permit the acquisition of 2D images from several slices in the moving sample.

3. ROTATIONAL RELATIVITY AND HALBACH NMR

3.1 Open-access Halbach NMR

Conventional NMR hardware encloses the sample in a nest of magnet, shim, gradient and RF coils which prevent easy access to the sample. In recent years considerable effort has therefore been spent on removing this restriction[2]. One-sided magnet systems, such as the NMR MOUSE, are open-access but are restricted to surface examinations. The 4-sided Halbach magnet array[8] (figure 8) is an attractive alternative that not only gives easy access to the sample but also probes the interior of whole samples. A recent feasibility study at the IFR in Norwich succeeded in measuring the T_2's of an intact hen's egg and a researcher's finger (still attached). The fact that the four magnets are well separated from the sample means that Halbach NMR is also well-suited to Rheo-NMR, where the sample is mechanically sheared or compressed during NMR acquisition. It should also be ideal for high pressure NMR and to applications where the sample is subjected to extremes of temperature. The later would, of course, require that each of the four magnets is enclosed in a thermostatically controlled cylindrical jacket.

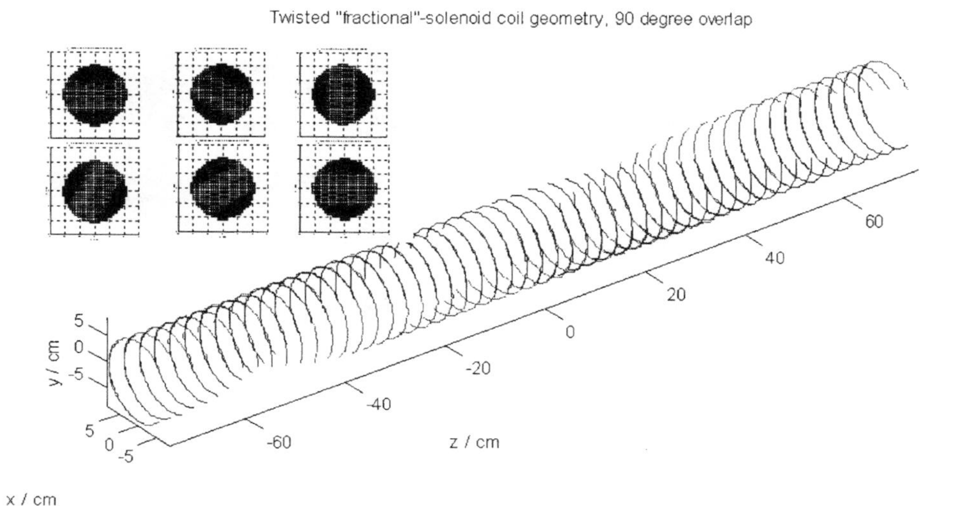

Figure 7 *The twisted helical transverse gradient coil for single-shot, constant current 2D COMSPIRA imaging. The inset shows the rotating gradient direction along the axis. The increasing turn density is not illustrated.*

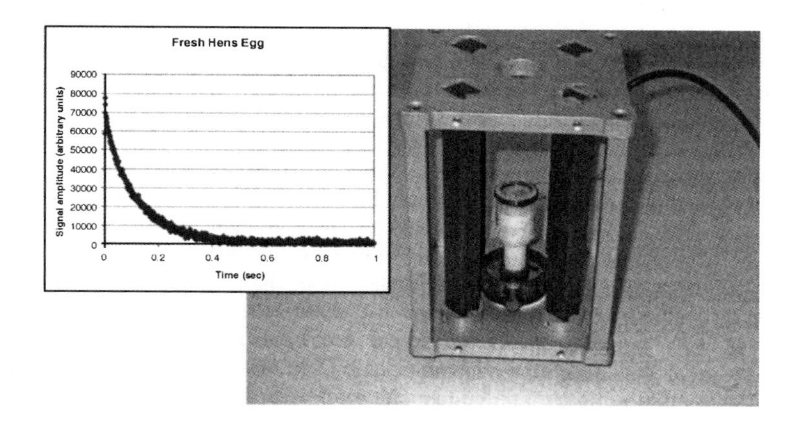

Figure 8 *The 4-magnet open-access Halbach array with the CPMG echo decay envelope of fresh hens egg (From ref. 8).*

So far the simple T_2 measurements have used a fixed array of the four magnets but the true potential of Halbach NMR emerges when the four magnets are allowed to rotate on their axes. Figure 9 shows that suitable coordinated magnet rotation can independently alter either the magnitude or the direction of the central magnetic field and this can, in principle, be used to develop Halbach field-cycling and RF-free NMR respectively.

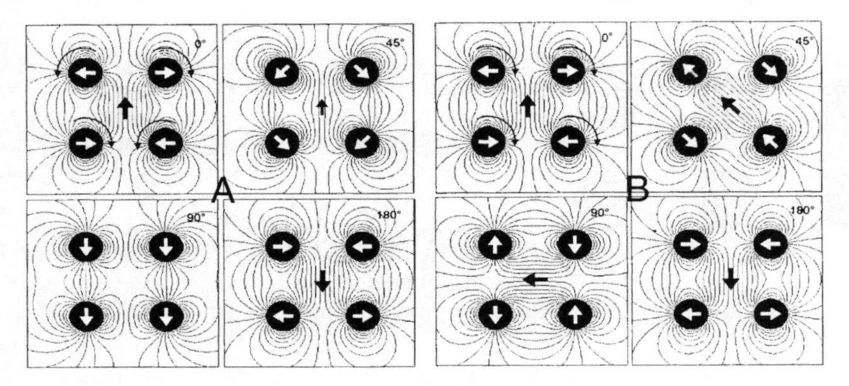

Figure 9 *The effect of A) counter-rotation and B) co-rotation on the central B_0 field in a 4-magnet Halbach array. Note how counter-rotation changes the magnitude and co-rotation the direction of the central magnetic field.*

3.2 Halbach Field-cycling NMR

Consider first the field-cycling aspects of Halbach arrays. Conventional field-cycling relaxometers reduce the field either by mechanically shuttling the sample in and out of the B_0 field or by reducing the current in an electromagnet. The former method is slow and suffers from mechanical vibration. The later method requires a very efficient cooling system to carry away the field energy. In contrast the Halbach system requires only that the four magnets are counter-rotated on their axes which can be done in a few milliseconds with pneumatic control. Finite element calculations show that the dependence of the central B_0 field on rotation angle can be reasonably well modelled as a simple cosine function which is ideal for field cycling applications because it means that inaccuracies in setting the rotation angle around the 0 or 180^0 regions will not greatly affect the magnitude of the maximum B_0 field used for polarisation and acquisition. The Halbach system also lends itself well to multidimensional field-cycling relaxometry. Bẏ acquiring a complete CPMG sequence after each field cycling $T_1(\omega)$ measurement it is clearly possible to measure a 3-dimensional $T_1(\omega)$-$T_2(\omega_0)$ spectrum where ω is the third dimension and ω_0 is the acquisition frequency. Such 3-d spectra could be particularly useful in separating overlapping peaks in the fixed frequency T_1-T_2 spectrum and, of course, provide simultaneous measurement of T_1 dispersion curves for each peak in the T_1-T_2 spectrum.

Further possibilities arise from the open-access nature of the 4-magnet Halbach array. NMR field-cycling is generally limited to a frequency range between a few kHz and a few tens of MHz, which contrasts with the huge frequency range available in dielectric relaxometry which goes from zero (DC) to about 10GHz. By incorporating a dielectric cell inside the Halbach array it should be possible to perform multidimensional, NMR-dielectric cross-correlation spectroscopy. Indeed the Halbach array should allow NMR field-cycling to be cross-correlated[9] with many other spectroscopic techniques such as FTIR, Raman or even X-ray scattering methods. This "simultaneous" acquisition of data from several spectroscopic techniques is especially valuable when the sample is undergoing rapid and irreversible changes as during food processing. Of course, to avoid electronic interference the dielectric and NMR acquisitions would not be truly "simultaneous" but rapidly alternated. Nevertheless Halbach NMR would appear to be ideal for developing such high-dimensional multispectral cross-correlation methods.

3.3 RF-Free Halbach NMR

Field cycling is perhaps the most obvious application of the Halbach system but it is not the only possibility. The fact that the direction of the B_0 field can be independently changed by co-rotation of the four magnets (see figure 9B) suggests the possibility of RF-free NMR in the sense that no RF excitation is required. Instead of creating transverse magnetisation and an FID by rotating the equilibrium longitudinal magnetisation with a 90^0 RF pulse, rotational relativity shows that the same can be achieved by leaving the magnetisation alone and rotating the external B_0 field through 90^0. Unfortunately this simple idea would not work in practice because the adiabatic principle states that the magnetisation would simply follow the changing direction of the external field unless the rate of change of the field direction is fast compared to the resonance frequency, ω_0. Because ω_0 is in the MHz range, impractically high magnet rotation velocities would be needed to perform non-adiabatic excitation. However the same result could, in principle, be achieved in a two-step process whereby the polarising B_0 field is first switched off by magnet counter-rotation in a time short compared to the samples T_1, then rapidly switched on in a perpendicular direction by reversing the direction of two magnets in opposite corners of the array (see figure 9). The resulting FID could then be detected in the usual way with a tuned and matched RF receiver coil. Spin-echoes could be created merely by reversing the B_0 field direction in the way suggested in figure 9. If such "RF-free Halbach NMR" succeeds it could form the basis of very low-cost off-line quality-control sensors based on simple FID, T_2 and T_1 measurements.

Size-rescaling of the Halbach array is yet another research avenue to be explored. Finite element simulations as well as simple dimensional arguments show that shrinking the size of the whole magnet array does not affect the strength of the central B_0 field. It should therefore be possible to develop capillary Halbach NMR suitable for micro-litre size samples and this would have the added bonus that the moments of inertia of the four smaller magnets would also be reduced, facilitating their rotation. Low-field multinuclear Halbach NMR might even be possible if the conventional RF coil is replaced with a SQUID detector.

4 CONCLUSION

Motional relativity clearly opens up many new and exciting possibilities for both on-line and off-line NMR. Both the translational on-line sensor and the rotational Halbach NMR sensor developments have been patented and are being researched and developed at the IFR in Norwich. Although the emphasis of this chapter has been on novel conceptual developments the technical challenges of designing and building the novel NMR hardware components are considerable and are, in fact, the rate determining step in the development. Although our experimental results[1,8] have established the feasibility of the simple one-dimensional acquisition protocols, the ultrafast higher dimensional developments discussed in section 2 and the rotational Halbach NMR applications discussed in section 3 are more speculative and remain to be tested. Nevertheless they serve to show the exciting potential of motional relativity in NMR.

Acknowledgments

The author (B.Hills) wishes to thank his dedicated research team for their tireless efforts. Thanks are also due to the BBSRC and DEFRA for funding support through their CASE studentship and LINK schemes respectively.

References

1. B.P.Hills and K.M.Wright, *J. Magn. Reson.*, 2006, **178**, 193.
2. B.P.Hills, *Annual Reports on NMR spectroscopy*, 2006, **58**, 177.
3. B.P.Hills and K.M.Wright, *Towards on-line NMR Sensors*, in Magnetic Resonance in Food Science, Ed. S.B.Engelsen, P.S.Belton and H.J.Jakobsen, Royal Society of Chemistry, Cambridge, 2005.
4. A.Tal, B.Shapira and L.Frydman, *J.Magn. Reson.*, 2005, **176**, 107.
5. N.Marigheto, S.Duarte and B.P.Hills, *Appl. Magn. Reson.*, 2005, **29**, 687.
6. N.Marigheto, K.M.Wright and B.P.Hills, *Appl. Magn. Reson.*, 2006, **30**, 13.
7. I.Rodriguez, R. Perez deAlejo, M.Cortijo and J.Ruiz-Cabello, *Concepts in Magn. Reson.*, 2004, **20B**, 40.
8. B.P.Hills, K.M.Wright and D.G.Gillies, *J.Magn. Reson.*, 2005, **175**, 336.
9. I.Noda and Y.Ozaki, *Two-dimensional Correlation Spectroscopy*, J.Wiley & Sons, New York, 2004.

MOLECULAR DYNAMICS IN SUGAR CLASSES AS REVEALD BY RECENT DYNAMIC SOLID-STATE NMR METHODS

D. Reichert[1], O. Pascui[1], M. Kovermann[1], N.E. Hunter[2] and P.S. Belton[2]

[1] Department of Physics, University of Halle, Friedemann-Bach-Platz 6, 06108, Halle, Germany
[2] School of Chemical Sciences and Pharmacy, University of East Anglia, Norwich, NR4 7TJ, UK

1 INTRODUCTION

Sugar glasses have found applications in encapsulation and stabilisation of labile therapeutic proteins and pharmaceuticals [1] and play a protective role in the dormant states of desiccation-resistant organisms [2]. In order to properly characterise these glasses there is a need to understand in detail the molecular dynamics of the systems. Nuclear magnetic resonance (NMR) offers a potentially useful tool and a number of studies have been published [3-7]. However there are some limitations: proton NMR relaxation time measurements are effectively limited to correlation time scales shorter than about 10^{-4}s due to the magnitude of the static dipolar interactions and the insensitivity of other relaxation mechanisms to motions slower than this. Similar considerations apply to deuterium studies although spin alignment methods [8;9] can in principle extend the dynamic range of the experiment towards slower motions, the necessity of isotopic labeling restricts its application for practical cases. 2D exchange experiments are attractive and have been described for the determination of rates of motion slightly above the glass transition temperature of glassy anhydrous glucose [4]. However these experiments are lengthy and require ^{13}C (or alternatively ^{2}H) enrichment. A potentially more convenient approach is (i) to make use of Magic-Angle Spinning experiments that improve the signal-to-noise and retain spectral resolution [10;11], making working with natural abundance samples feasible and (ii) use 1D variants of the exchange NMR methods. These two approaches were combined to the 1D trODESSA experiment [12] and later developed into the CODEX method [13;14], which is a robust and time saving experiment that permits the quantitative determination of both time scale and amplitude of molecular dynamic processes with the molecular resolution of a Solid-State ^{13}C spectrum in natural abundance organic solids. Principles and applications for synthetic polymers [15] and solid proteins [16] can be found in recent review articles.

In this paper, we introduce the NMR method, discuss opportunities and limitations and present its application to glassy Methyl α-L-Rhamnopyranoside (Me-Rha). Since Exchange-NMR methods are not abundantly used so far in the study of Carbohydrates, we discuss the Exchange NMR in general, its classification into the dynamic range, the extension to MAS and 1D and its information content in some detail.

2 METHODS AND MATERIALS

1.1 MAS-Exchange NMR: CODEX

The time scale of molecular dynamic processes spans many orders of magnitude and can be investigated by a number of experimental techniques. Among them, mechanical and dielectric relaxation are the most prominent one. However, they suffer from the drawback that the experimentally detected dynamics modes cannot easily be assigned to the molecular structures. For example, the dielectric relaxation detects the reorientation of the molecular dipolar moment, i.e. it basically tells only about the mobility of this part of the molecule. This makes it difficult to tell apart different dynamic processes originating from different structures in the molecules (side-chains in polymers, exocyclic groups in carbohydrates etc.). Solid-State NMR, on the other hand, is able to deliver information about both the time scale of motion (correlation time τ_c) as well as information about the amplitude of motion (jump angles etc.) over about 12 decades of the correlation time (Figure 1).

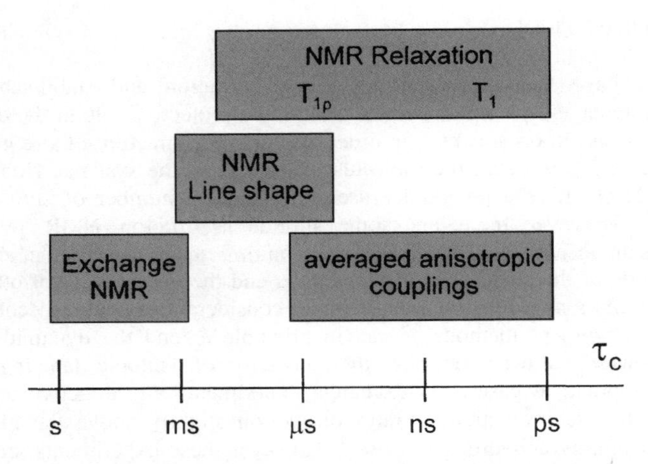

Figure 1 *Dynamic Range of different Dynamic Solid-State NMR Methods*

Of particular interest are slow molecular processes with correlation times in the order of milliseconds, since they relate to the mechanical properties of the material and play important roles in food-relevant processes such as aging. These processes can be investigated by Exchange -NMR methods [17-19] the make use of the orientation dependence of the NMR resonance frequency in solids: the orientations of the probe nucleus before and after and adjustable so-called mixing period τ_m are being compared via the resonance frequencies in the schema of a a two-dimensional experiment (Figure 2). A 2D data set is acquired and a 2D-FT yields a 2D-Exchange spectrum, in which spectral features off the diagonal (cross peaks), i.e. at $\omega_1 \neq \omega_2$ tell about a molecular dynamic process, realized by change of orientation of the probe nucleus during τ_m. The correlation time of motion can be determined by running a series of experiments with increasing value of τ_c and fitting the growth of cross peaks to an kinetic equation with in the easiest case of a symmetric two-site jump becomes an exponential function[20].

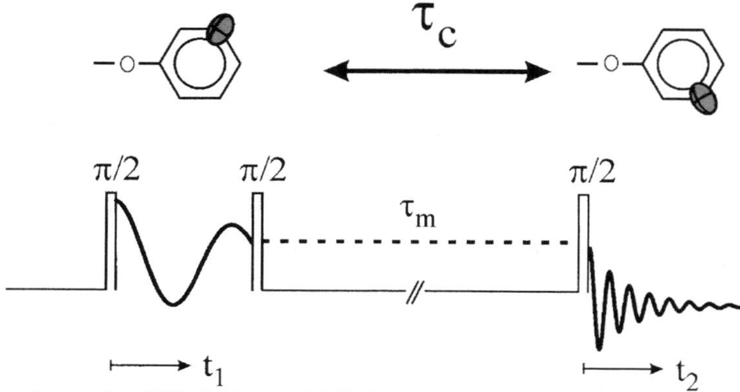

Figure 2 *Principle of 2D-Exchange NMR Spectroscopy*

The obvious nucleus to be observed in organic solids is the isotope ^{13}C with a natural abundance of 1.1%. Each molecule has a number of chemically inequivalent ^{13}C nuclei that produce different resonances in the NMR spectrum, providing the opportunity to retrieve dynamic information from different molecular parts. However, these resonances normally overlap in the Solid-State NMR spectrum and the low natural abundance causes additional detection problems. The experimental measure to overcome these problems is the application of Magic-Angle Spinning, MAS[21], however, the two-dimensional data acquisition still requires long experimental times.

The way to shorten the acquisition time but to keep the information content of the NMR data comes from the idea that it is not necessary for the detection of a frequency to acquire the complete NMR signal in t_1 by incrementing this time parameter in a two-dimensional fashion but to keep t_1 constant and measure the phase that the magnetization has acquired before and that is reconverted after the mixing period. In the case of no molecular dynamic process, the resonance frequencies of the observed nucleus and thus the phases acquired by the magnetization before and after τ_m are the same and a full signal is being acquired while any molecular reorientation leads due to incomplete compensation of de- and rephrasing to an attenuation of the detected NMR signal. Thus, the dynamic information that is mapped in the 2D spectrum in an experiment as shown in Figure 2 is now partially converted into line intensities of a 1D spectrum. The correlation time of motion τ_c can detected by rung a series of time-saving 1D Exchange experiments for different values of τ_m. Such experiments were the ODESSA (One-dimensional Exchange[22] Spectroscopy by Sides-band Alternation), the trODESSA (time-reversed ODESSA) [12] and the CODEX (Centerband only Detection of Exchange) [13;14]. The latter provides the additional possibility to determine the amplitude of the molecular process. For that, the experiment can be performed in two modes (Figure 3): For the determination of τ_c , the length of τ_m is being incremented and the resulting decay is fitted to an kinetic equation. It should be noted that this decay is proportional to the correlation function of motion so not only the correlation time but also the shape can be investigated. For amorphous materials, the correlation function often deviated from a single exponential decay and can for all practical purposes be described by a stretched- exponential or KWW function[23]

$$\frac{S}{S_0}(\tau_m) = (1-a) + a e^{-\left[\frac{\tau_m}{\tau_c}\right]^{\beta}} \quad (1)$$

the parameters of which are the correlation time τ_c , the width parameter β and the plateau value *a*, see Figure 3, left. To determine the amplitude of motion, however, the length of the mixing period τ_m is kept constant and the lengths of the so-called recoupling periods before and after τ_m are being incremented. In the CODEX experiment, this can be done by either varying the MAS-spinning speed $1/T_R$ or to add more rotor-synchronized evolution periods, *N*, in which π-pulses enhance the dephasing effect of the magnetization. The larger the amplitude of motion, the faster the signal decays to an asymptotic value. Information about the motional amplitudes (jump angles) can be extracted from a comparison of the acquired decay calculated data and from the value of the asymptotic plateau [14].

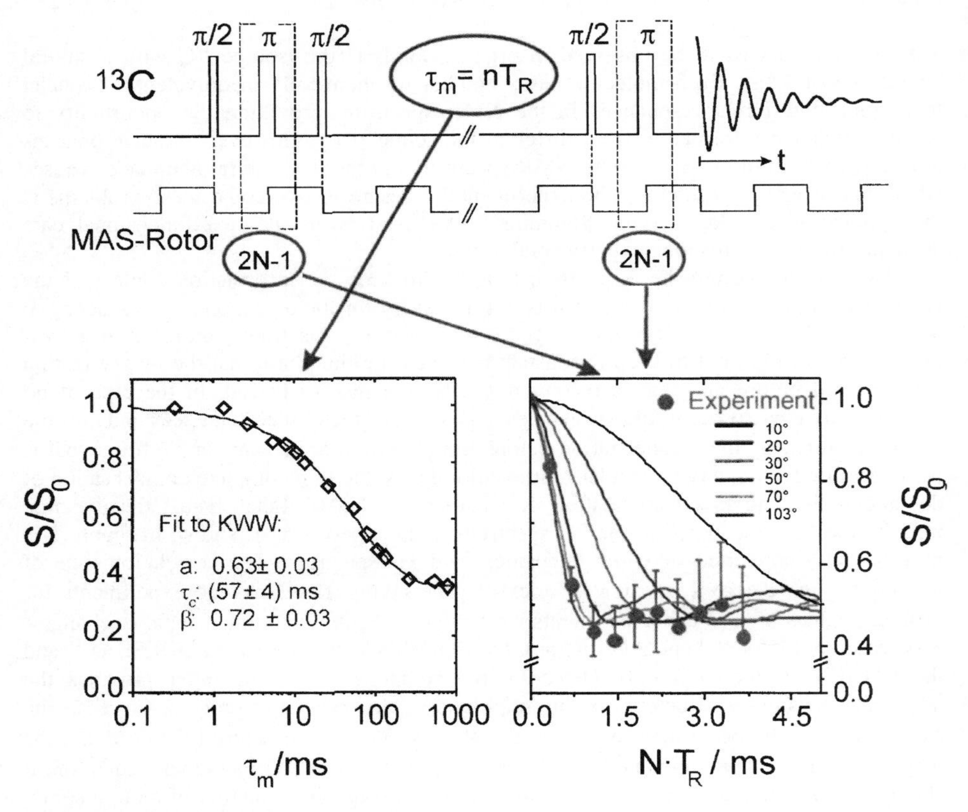

Figure 3 *Principle of the CODEX 1D-MAS Exchange experiment. Experimental and simulated data were taken from*[24].

Additional features of the experimental setup are the acquisition of a normalized signal S/S0 and the suppression of spin-exchange effects. Details are discussed in [14;25;26].

1.2 Methyl α-L-Rhamnopyranoside

Methyl α-L-Rhamnopyranoside (Me-Rha) is an important synthetic intermediate and a partial structure of a growing family of pyranonaphthoquinone antibiotics[27-29] which have shown significant antimicrobial and potential antitumor activities[27;29]. Here we present preliminary results on a sample of this material. (Me–Rha) was purchased from Sigma and used as the starting materials without further purification. The authenticity of these samples was ascertained by ^{13}C CPMAS spectra and melting point measurements. The T_g is reported to be 290K [30]. In order to make the glass, we heated a crystalline sample in a container and hold it for about twenty minutes at 390K and repeated the heating and cooling cycle three times.

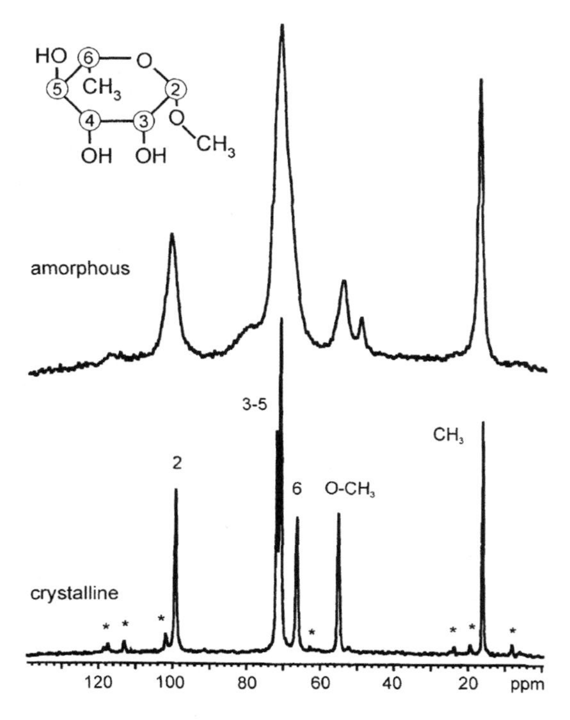

Figure 4 *CPMAS spectra of Me-Rha.in the amorphous and crystalline state. (*) denote spinning sidebands from the MAS experiment.*

3 RESULTS AND DISCUSSION

Figure 4 shows ^{13}C-MAS spectra for the crystalline and amorphous sample. Typical acquisition parameters are a MAS rate of 3 to 5kHz, a repetition delay of 5s and between 256 and 4k accumulations. The ^{13}C line assignment is added to the spectrum of the crystalline sample. Turning into the amorphous structure causes severe line broadening as a result of the lost translational symmetry of the molecular packing. The additional peak at 48ppm in the amorphous spectrum is very likely due to methanol that was created during the preparation procedure. This may have an effect on the detailed behaviour of the material and points to the fact that great care is need in the preparation of these samples if they are to be pure. Much of the literature does not attempt to characterise the material after heating except to note its glass transition temperature, so the actual purity of the samples is unknown

For the quantitative determination of the amplitude of motion, the value of the Chemical-Shift anisotropy, CSA, is needed as an input parameter for the calculation of the NT_R dependence (Figure 3, right) and has to be determined experimentally. This can be done by the analysis of the spinning-sideband, ssb, patterns of the individual resonance at a relatively low MAS-spinning speed[11]. This, however, leads to a crowded spectrum as shown in Figure 5, bottom. Special 2D pulse sequences disentangle the ssb's belonging to different carbons. The PASS experiment[31] proved being a robust approach to separate the ssb patterns from different carbon resonances according to their order. Figure 5, top, shows the PASS spectrum of crystalline Me-Rha recorded at a spinning rate of 1kHz. Applying the procedure of [11] yield CSA values of about 45ppm for the CH carbons, 72ppm for O-CH_3 and 29ppm for CH_3. The asymmetry parameters for CH_3 and O-CH_3 are close to 0. Only these carbons are used in the analysis of the amplitude of motion, see below.

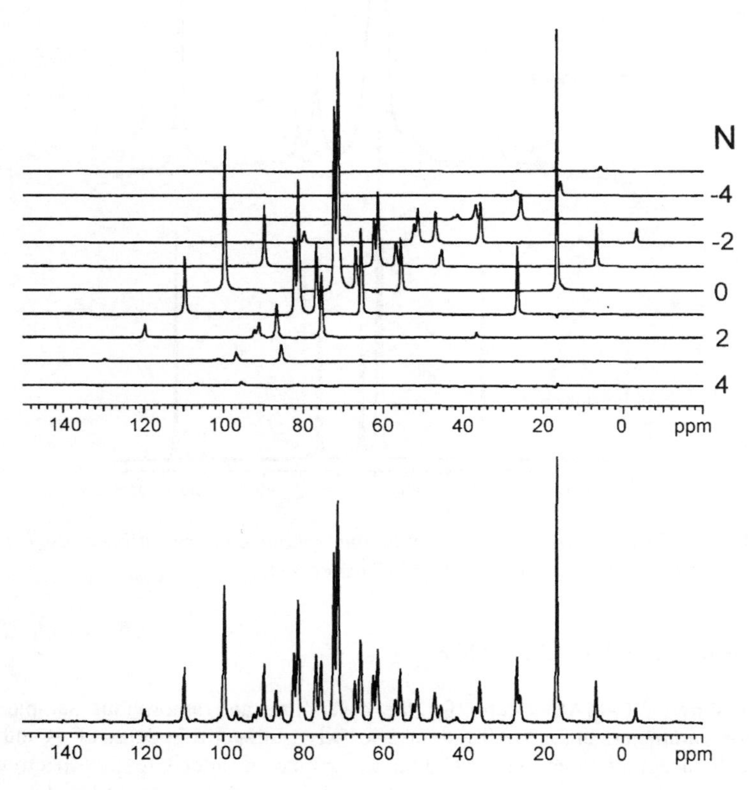

Figure 5 *MAS (bottom) and PASS spectra (top) of Me-Rha, taken at a MAS rate of 1 kHz. The MAS spectrum is the superposition of all slices of the PASS spectrum.*

Figure 6 shows CODEX τ_m dependencies for different carbons in glassy Me-Rha, taken at a MAS rate of 4 kHz, N=2 and at different temperatures. The full lines are fittings to Eq.(1) . The extracted correlation times are listed in Table 1 while the β values varies around 0.5±0.1, indicating a distribution of correlation times of about 1.5 decades. In contrast, the crystalline sample does not exhibit any dynamic process in the chosen temperature interval and within the dynamic range of the CODEX experiment. This implies that the motion is too slow to be observed in this experiment. The correlation times for the ring carbons and the CH_3 are similar while the O-CH_3 exhibits a slightly shorter correlation time, in particular at the highest temperature. This might indicate an additional process of the exocyclic O-CH_3 group that cannot be separated from the reorientation of the entire molecule in these data.

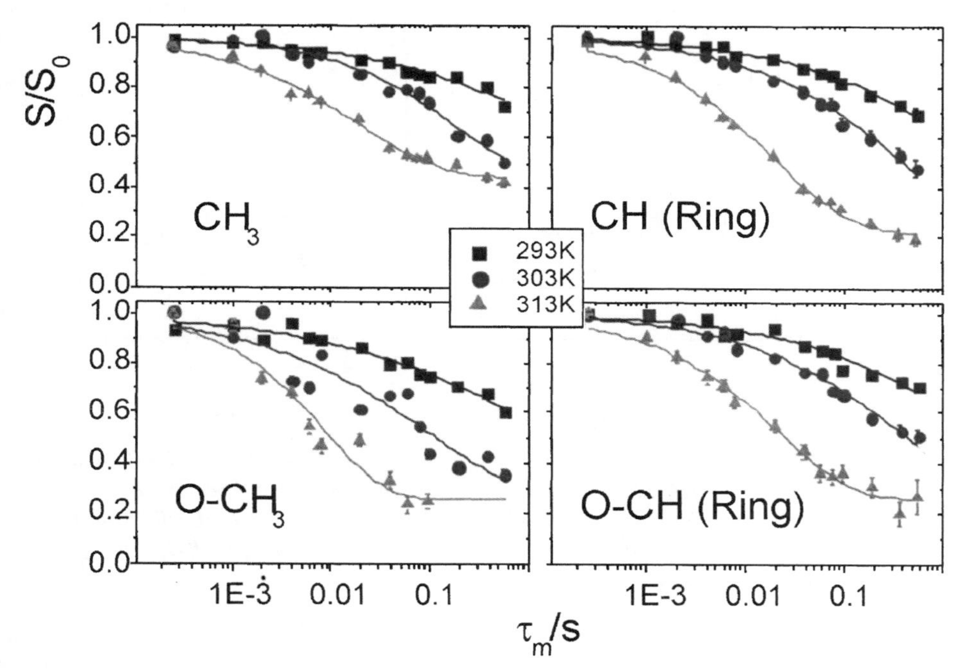

Figure 6 *CODEX τ_m dependencies of different resolved peaks of glassy Me-Rha at different temperatures. The MAS rate was 4 kHz and N=2 were used.*

Additional information can be extracted from experiments aimed to the amplitude of motion (NT_R dependence of the CODEX experiment). Figure 7 show the data taken at τ_m =100 ms and 313 K. The left hand side shows the data as acquired. Lines connecting the points serve to guide the eye. However, these curves not only depend on the amplitude of motion but also on the value of the CSA. Then right hand side thus displays the data scaled with the CSA values that where extracted from the data shown in Figure 5. It suggests that the ring and the CH_3 group are exposed to the overall motion of the molecule while the O-CH_3 group performs an additional reorientation, due to the additional degree of freedom around the C-O bond. A quantitative evaluation requires not only the knowledge of the

CSA values but also the orientation of the CSA tensor in the molecule. This is well known for CH_3 carbon in which the tensor is in good approximation axially symmetric and oriented along the C-C bond. For that reason, only the CH_3 and O-CH_3 carbon are considered. Comparison with calculated data yield jump angles of 15° for the CH_3 group (and thus for the entire molecule) and 30° for the O-CH_3 group, confirming an additional reorientational process. The asymptotic plateau of about 0.2 excludes a rotational-diffusion like process but suggest a jump process between a limited number of sites.

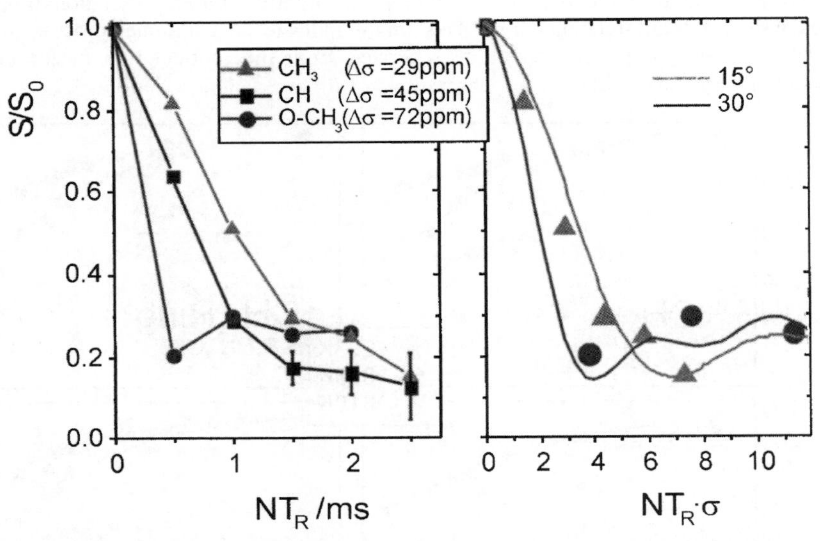

Figure 7 *CODEX NT_R dependencies of different resolved peaks of glassy Me-Rha at 313. The MAS rate was 4 kHz and τ_m=100ms were used. The data on the right side are taken from these displayed on the left side and scaled with the different CSA values. Full lines at the right hand side are calculations based on the given jump angles.*

Table 1 *Correlation times of motion derived from the data shown in Figure 6, in milliseconds. The error margin is about 20%.*

	293 K	303 K	313 K
CH (Ring)	2800	443	22
CH_3	2100	195	21
O-CH_3	1500	90	8

4 SUMMARY

We have demonstrated the first application of Solid-State 1D-MAS exchange NMR experiments for the investigation of slow reorientational processes in glassy Carbohydrates. While the crystalline samples have reorientation rates that are too slow to be observed in the temperature range studied the molecules in the glassy sample exhibit

jump like reorientations with amplitudes of about 15° and with correlation times in the order of milliseconds at temperatures above T_g. The exocyclic O-CH_3 group performs an additional jump process with similar correlation times.
Whilst these results must be considered to be provisional because of the problem of decomposition encountered they do demonstrate the potential of the method for measuring slow motions in soft solids.

5. ACKNOWLEDGEMENT

The authors thank the BBSRC and Pharmorphix Ltd for financial support.

References

1 Yu, L., *Advanced Drug Delivery Reviews*, 2001, **48**, 27.
2 Crowe, J. H., Carpenter, J. F., Crowe, L. M., *Annual Review of Physiology*, 1998, **60**, 73.
3 Hills, B. P., Pardoe, K., *Journal of Molecular Liquids*, 1995, **63**, 229.
4 van Dusschoten, D., Tracht, U., Heuer, A., Spiess, H. W., *Journal of Physical Chemistry A*, 1999, **103**, 8359.
5 Moran, G. R., Jeffrey, K. R., *Journal of Chemical Physics*, 1999, **110**, 3472.
6 Hills, B. P., Wang, Y. L., Tang, H. R., *Molecular Physics*, 2001, **99**, 1679.
7 Tang, H. R., Wang, Y. L., Belton, P. S., *Physical Chemistry Chemical Physics*, 2004, **6**, 3694.
8 Spiess, H. W., *Journal of Chemical Physics*, 1980, **72**, 6755.
9 Lausch, M., Spiess, H. W., *Journal of Magnetic Resonance*, 1983, **54**, 466.
10 Andrew, E. R., Bradbury, A., Eades, R. G., *Nature*, 1958, **182**, 1659.
11 Herzfeld, J., Berger, A. E., *Journal of Chemical Physics*, 1980, **73**, 6021.
12 Reichert, D., Zimmermann, H., Tekely, P., Poupko, R., Luz, Z., *Journal of Magnetic Resonance*, 1997 , **125**, 245.
13 deAzevedo, E. R., Hu, W. G., Bonagamba, T. J., Schmidt-Rohr, K., *Journal of the American Chemical Society*, 1999, **121**, 8411.
14 deAzevedo, E. R., Hu, W. G., Bonagamba, T. J., Schmidt-Rohr, K., *Journal of Chemical Physics*, 2000 , **112**, 8988.
15 deAzevedo, E. R., Bonagamba, T. J., Reichert, D., *Progress in Nuclear Magnetic Resonance Spectroscopy*, 2005, **47**, 137.
16 Krushelnitsky, A., Reichert, D., *Progress in Nuclear Magnetic Resonance Spectroscopy*, 2005, **47**, 1.
17 Jeener, J., Meier, B. H., Bachmann, P., Ernst, R. R., *Journal Of Chemical Physics*, 1979, **71**, 4546.
18 Edzes, H. T., Bernards, J. P. C., *Journal Of The American Chemical Society*, 1984, **106**, 1515.
19 Schmidt, C., Wefing, S., Blumich, B., Spiess, H. W., *Chemical Physics Letters*, 1986, **130**, 84.
20 Luz, Z., Tekely, P., Reichert, D., *Progress in Nuclear Magnetic Resonance Spectroscopy*, 2002, **41**, 83.
21 Dejong, A. F., Kentgens, A. P. M., Veeman, W. S., *Chemical Physics Letters*, 1984, **109** , 337.

22 Gerardymontouillout, V., Malveau, C., Tekely, P., Olender, Z., Luz, Z., *Journal of Magnetic Resonance Series A*, 1996, **123**, 7.
23 Williams, G., Watts, D. C., *Transactions Of The Faraday Society*, 1970, **66**, 80.
24 Miyoshi, T., Hayashi, S., Imashiro, F., Kaito, A., *Macromolecules*, 2002, **35**, 2624.
25 Reichert, D., Bonagamba, T. J., Schmidt-Rohr, K., *Journal of Magnetic Resonance*, 2001, **151**, 129.
26 Pascui, O. F., Reichert, D., *Applied Magnetic Resonance*, 2004, **27**, 419.
27 Brimble, M. A., Duncalf, L. J., Nairn, M. R., *Natural Product Reports*, 1999, **16**, 267.
28 Brimble, M. A., Nairn, M. R., Prabaharan, H., *Tetrahedron*, 2000, **56**, 1937.
29 Krishnan, P., Bastow, K. F., *Cancer Chemotherapy and Pharmacology*, 2001, **47**, 187.
30 Tang, H. R., Wang, Y. L., Belton, P. S., *Physical Chemistry Chemical Physics*, 2004, **6**, 3694.
31 Antzutkin, O. N., Shekar, S. C., Levitt, M. H., *Journal Of Magnetic Resonance Series A*, 1995, **115**, 7.

HOW MUCH INFORMATION IS THERE IN AN NMR MEASUREMENT?

P.S. Belton

School of Chemical Sciences and Pharmacy, University of East Anglia,
Norwich NR4 7TJ, UK

1 INTRODUCTION

In many applications of analysis the problem is to classify samples into various sets. Typical of these types of analytical problems are authentication and metabolomics. The analytical results on the sample must be compared with the members of different sets. In many cases NMR is the analytical method of choice, however often the choice of analytical method is determined by practical considerations such as equipment availability, the skills of the analyst and cost. A method is then developed which may or may not solve the problem. Instead of going through the lengthy trial and error processes is it possible to make some choices a priori about what methods are likely to be more successful on the basis of an objective analysis?

An approach to this problem is to consider the information content of the analytical method and compare with the information requirements of define the membership of a set. The amount of information in the spectrum can then be compared with that required to define membership of a set. Information is a term that is used in different ways. In normal conversation it is often taken a synonymous with meaning. However this definition, whilst convenient, is far from universally applicable. Any one who has struggled trying to understand a foreign language becomes immediately aware that information, in the form of the language in which one is being addressed, is plentiful but meaning is lacking. Similarly if one cannot read musical notation the information in a musical score is meaningless. This led Weaver[1] to derive a definition of information that did not depend on meaning:

> "The word *information*, in this theory, is used in a special sense that must not be confused with its ordinary usage. In particular, information must not be confused with meaning. In fact, two messages, one of which is heavily loaded with meaning and the other of which is pure nonsense, can be exactly equivalent, from the present viewpoint, as regards information."

Information is therefore not to be confused with meaning. Indeed meaning can only be derived when there is some additional knowledge that exists externally to the information and is held by the interpreter, as illustrated by the examples given above. Information, as opposed to meaning, has the advantage that it is unbiased in the sense that all data is information and equally valid regardless of the knowledge or the implicit assumptions of the receiver.

2 THE INFORMATION IN A SPECTRUM

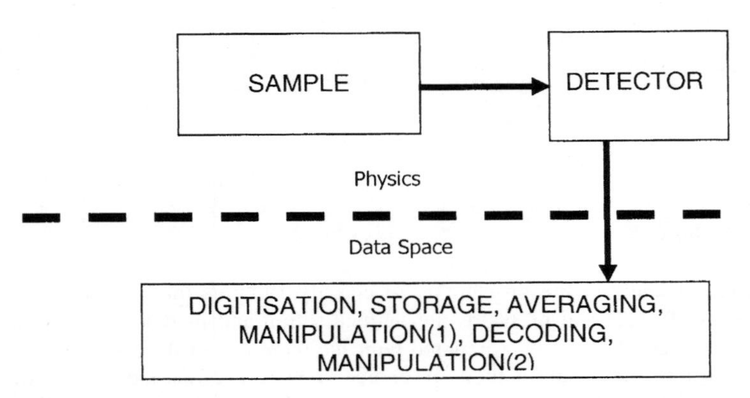

Figure 1 *Schematic representation of an NMR spectrometer*

Figure 1 is a representation of an NMR spectrometer. The sample is the source of signal induced in a coil, which is detected as a time variant voltage. This signal, which is always mixed from noise from a variety of sources, after amplification and other forms of manipulation, which may themselves generate additional noise, is digitised and further manipulated, typically by some sort of additive averaging, which may involve scaling operations. Often digital filtering and apodisation is applied before Fourier transformation and afterwards other manipulations such as phase correction and base line correction are carried out. The point at which digitisation takes place is critical one, it represents the transition between purely physical process and data space. In data space there only exist ordered collections of numbers. Inferences may be made about physical processes from these. The numbers themselves do not contain meaning, only information. Meaning comes from inferences drawn from them with the application of prior knowledge.
Shannon[2] proposed a way of quantifying the information content of a message by defined a quantity, ρ_i, which was the probability of the ith message. In terms of spectrum this may be equated to the normalised intensity of the data point. For a discussion Shannon's concepts with respect to NMR see the discussion by Wright[3].

Shannon defined quantity called the entropy of the message as

$$S = -\sum_i \rho_i \log_2 \rho_i \qquad (1)$$

For a spectrum, this needs to be redefined as

$$S_s = -\sum_i P_i Ln P_i \qquad (2)$$

Where,

$$P_i = \frac{I_i}{\sum I_i} \qquad (3)$$

S is the Shannon entropy. In equation the logarithm is changed to the base e for convenience. I_i is the intensity of the ith point. P_i is termed the relative intensity of ith point The entropy, like the thermodynamic entropy, is a measure of spreading, in this case how spread out the data is across the spectrum. This is illustrated in figure 2. The high entropy spectrum has its intensity more evenly distributed than the low entropy spectrum.

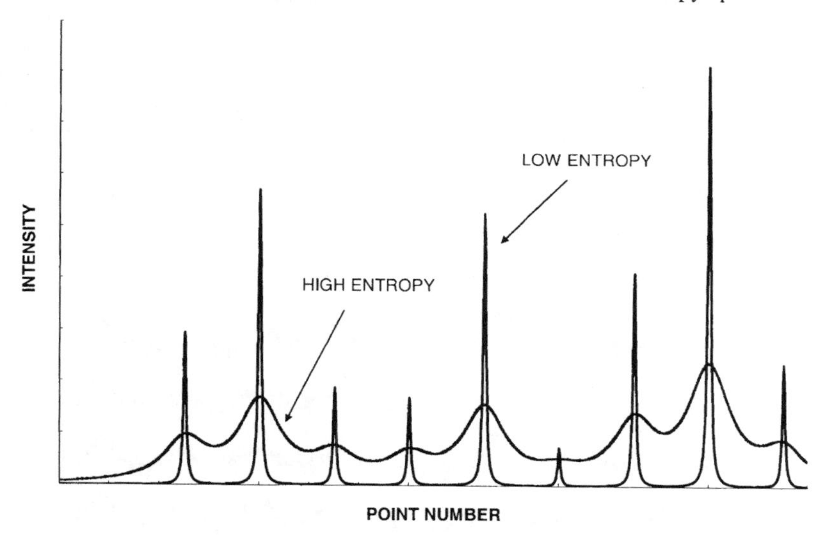

Figure 2 *Spectra showing low and high entropy*

The entropy as defined here has some important properties: the term P_i is always less than one so that the logarithmic term is negative hence the negative sign in front of the summation. P_i cannot take a negative value or be zero. This appears to be problem since the NMR spectruni is intrinsically a complex quantity and thus may have negative and positive values and can be zero. However a sign attached to a number is a second piece of information, in data space as opposed to physical space this corresponds to a direction of a deviation from an arbitrary zero. The problem can safely be dealt with by simply adding a large arbitrary value to all the data points so that none are negative or zero. This does not affect the information content of the data materially, although if the data is analysed

without the analyst being aware of the addition it may affect the interpretation. This is another illustration of the distinction between meaning and information.
A further important point about the entropy is that it is an extensive variable. This means that the more data points that are available the greater the information content. This leads to the apparently anomalous potion that an operator can increase the information content of a spectrum simply by zero filing the FID before Fourier transformation. This is true. The information in the spectrum is increased but the cause of the increase is simply the additional information added by the operator. The source of the information is irrelevant to the information content but it is of critical importance in ascribing meaning to the spectrum.

3 THE INFORMATION IN A SET OF SPECTRA

Typically in an application, samples will be sorted into sets for classification purposes. The data may be expressed as a set of spectra plotted out or as a matrix. Generally it is required that the spectra obtained within a set are similar. If they are not similar then it will not be possible to differentiate between the one set and another set. Similarity between spectra implies that any point in the nth spectrum of set with a relative intensities $P_{i,n}$ will have a similar value to the points $P_{i,n}$ for all n. A graph of the values for $P_{i,n}$ versus n for a given i will therefore tend to flatness. This is condition of high entropy. Since in the data matrix n defines the columns of the matrix and i defines the rows, a requirement for similar spectra within a set is that the column entropy is high. This is illustrated in figure 3.

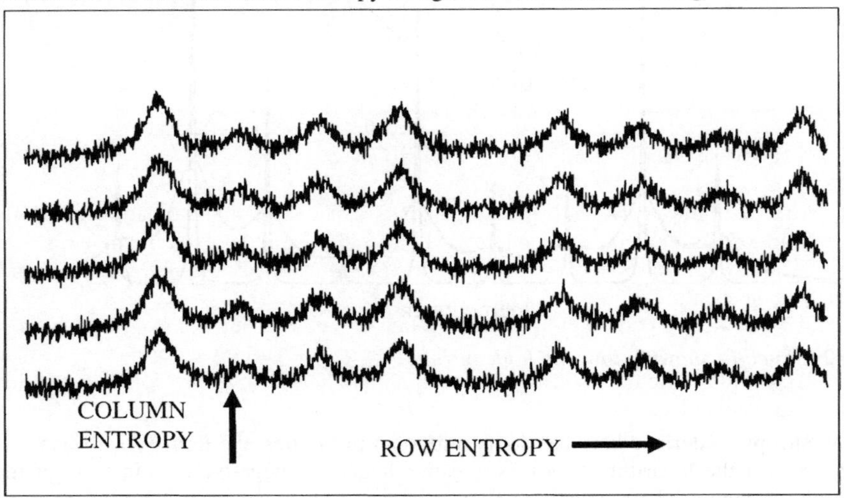

Figure 3 *Illustration of a set of spectra showing the directions in which row and column entropy are calculated*

The highest values of column entropy will result from the highest similarity of all the spectra in the set. However the highest values of row entropy will result in a flat spectrum with no features. This is unlikely to contain useful data. In order to be useful the spectrum must have some structure. An additional criterion to the entropy criterion is therefore needed to ensure that the spectrum has structure. A useable definition of structure is:

$$\frac{\sqrt{\sum_i \left(I_i - \bar{I}\right)^2}}{N-1} > \sigma_n \tag{4}$$

Where $\bar{I}$ the mean value of *Ii*, N is is the number of data points in the spectrum and σ_n is the standard deviation of the noise.

The use of this approach implies some prior knowledge about the noise in the system and therefore by implication a definition of noise and signal. On a purely information basis these distinctions cannot be made.

In figure 4 there is an illustration of how entropy sums up across a spectrum. Each value of entropy plotted represents the entropy of the spectrum up to that point. Note that the noise contributes to the entropy since it is information but that small peaks contribute much less than large peaks.

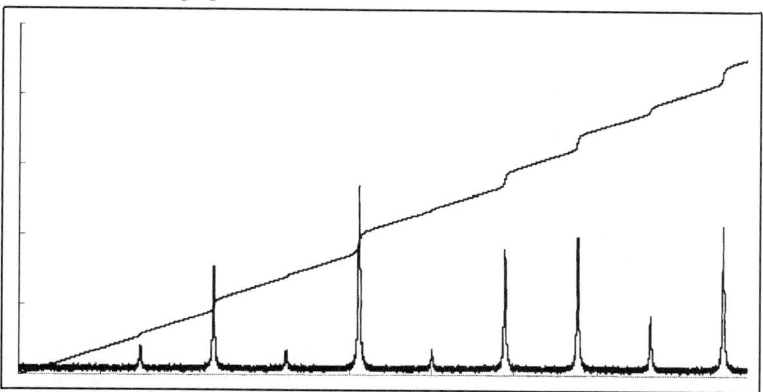

Figure 4 *The variation of entropy across a spectrum.*

The relative contribution of small and large peaks can distort the estimation of useful information in a spectrum, since in modern NMR spectrometers the dynamic range is very high and generally the noise level is very small. This is illustrated in figure 5 taken from a 600 MHz proton spectrum of a beer sample[4].

The signals from the aromatic region clearly contain useful data but would contribute only marginally to the entropy. In order to better express the useful information content therefore a new definition of entropy is required that takes account of the distinction between signal and noise.

A way of doing this is redefine the intensity I as I'such that for

$$I_i^2 > \sigma_n^2,\ I_i' = 1{+}1,\ \text{else},\ 1{+}0 \tag{5}$$

Re-applying equation 2 then gives a measure of the useful information content in the spectrum.

Thus:

$$S' = \sum_i P_i' \ln P_i' \tag{6}$$

$$\text{Where } P_i' = \frac{I_i'}{\sum_i I_i'} \tag{7}$$

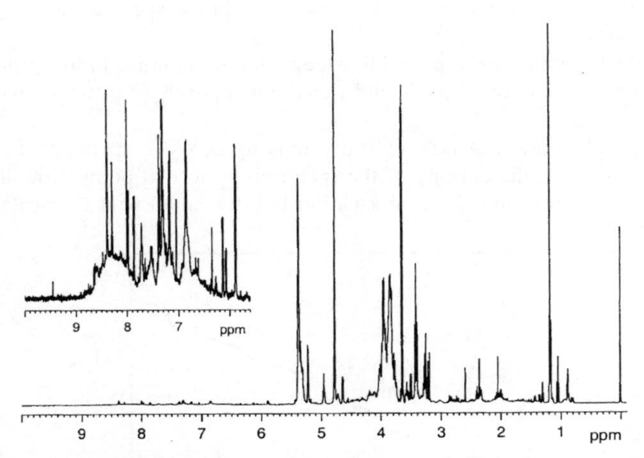

Figure 5 *Proton NMR spectrum of a beer sample. Insert shows a magnification of the aromatic region*

4 COMPARING SPECTRAL SETS

In many analytical applications the problem is to classify the sample under test as member of a set. In an application such as metabolomics most sets will meet the criteria set out above: they will have high row and column entropy and meet the structure criterion set out in equation 4. However in the case of authentication, for example, the authentic set will contain a finite number of similar spectra. The spectra in inauthentic set may vary without limit, may have some features in common with the authentic set but must also have features different from the authentic set. In this case there are no restrictions on the entropy of the inauthentic set. In order to compare spectra within the authentic an inauthentic set it is necessary to define the probability, $\mathbb{P}$, that an intensity $I_{i,a}$, which is the intensity of the ith point in the authentic set and the intensity in the inauthentic set, $I_{i,\mu}$ are equal. More generally $\mathbb{P}$ will represent the overlap of the volumes occupied by the all values of I_i in suitably dimensioned spectral space. $\mathbb{P}$ can take values between 0 and 1, best discrimination is when $\mathbb{P}$ is very much less than one. Therefore we require that:

$$\mathbb{P}(I_{i,a} = I_{i,\mu}) << 1 \tag{8}$$

Since the spectra contain n data points. The total probability of the spectra being different is given by:

$$\Pi_n \mathbb{P}(I_{i,a} = I_{i,\mu}) << 1 \tag{9}$$

Where

$$\Pi_n = P(I_{1,a} = I_{1,\mu}) * P(I_{2,a} = I_{2,\mu}) \ldots$$

There are two ways in which condition 9 may be met. If n is small then each value of P must be much less than 1. If, on the other hand, n is large then each value of P can be very close to one but the product will still be much less than one.

In the case of SNIF NMR for deuterium isotope measurements[5] the spectrum is very simple, only two peaks are important: the CD_2 and CD_3 peaks of ethanol. The spectra in the authentic set therefore have low row entropy. Since they are to be compared with possibly inauthentic samples that have only slightly different intensities the column entropy must be very high. That is all the spectra of samples within the authentic set must be very similar. Since there relatively few peaks in the spectrum the value of n is also small. Under these circumstances the condition that $P(I_{i,a} = I_{i,\mu}) << 1$ must be met for all data points.

In the case of a 1D, high field, proton spectrum such as is obtained for a biological fluid, the spectrum is richly populated and the row entropy is high. The column entropy will typically be lower than in the SNIF case but this is compensated by the large value of n. in this case condition 8 can be relaxed since condition 9 will be met by the multiplication of the individual values of P.

The use of SNIF NMR depends on the understanding of the significance of the information in the spectrum and a knowledge of the likely behaviour of the set of inauthentic samples. The actual samples themselves are the product of a very well controlled preparation process to ensure that no interfering substances are present and isotope ratios are maintained. In this way the apparently sparse information content of the spectra are compensated by a considerable amount of *a priori* knowledge.
In contrast to this, 1D proton spectra are often used with minimal preparation and in systems which are much less than completely characterised. In this case the high information content of the spectra compensates for the lack of *a priori* knowledge.

Generalising these examples it may be said that:
Where there is sufficient a priori knowledge low row entropy methods may be used

- Where the level of a priori knowledge is low, high row entropy methods should be used
- In both cases the highest possible column entropy is desirable.

References

1 C. Shannon and W. Weaver, *The Mathematical Theory of Communication,* University of Illinois Press, Urbana, 1964, p 8
2 C. Shannon and W. Weaver, *The Mathematical Theory of Communication,* University of Illinois Press, Urbana, 1964, p 50
3 K Wright 'Maximum Entropy Methods in NMR Data Processing' in *Signal Treatment and Signal Analysis in NMR,* ed. D. N Rutledge, Elsevier, Amsterdam, 1966, Chapter 2, pp 25-43
4 I. Duarte, A. Barros, P. Belton, R .Righelato, M. Spraul, E. Humpfer and A. Gil, *J. Agric. Food Chem.*, 2002, **50**, 2475
5 G Martin 'Advances in the Authentication of Food by SNIF NMR' in *Magnetic Resonance in Food Science*, eds. S. B. Engelsen, P. S. Belton and H. J. Jackobsen, Royal Society of Chemistry, Cambridge, 2005, pp 31-38

ADVANCES IN THE MAGNETIC RESONANCE IMAGING OF EXTRACELLULAR MATRIX OF MEAT

J.M. Bonny, L. Foucat, M. Mouaddab, L. Sifre-Maunier, A. Listrat and J.P. Renou

QuaPA – STIM, INRA Theix, 63122 Saint Genes Champanelle, FRANCE

1 INTRODUCTION

The extracellular matrix (ECM) - also named intra-muscular connective tissue – has several important purposes for the muscle. It binds fibres in a complex multi-scale arrangement of fascicles, guides the branches of nerves and blood vessels through the muscle, stores fat and fills extracellular space. ECM is an inhomogenously distributed overlap of different compounds (i.e. lipid, collagens, decorin, elastin). The properties of each compound are implicated in food science for understanding the inherent biological factors that determine the textural properties of both meat and fish flesh. In meat, toughness is the main problem, while in fish, splitting of the flesh is the main issue.
Magnetic resonance imaging (MRI) is relevant for the non-destructive analysis of ECM because it has potential for giving information on both the spatial distribution and the composition. Up to now MRI of ECM has laid little attention, except for imaging intramuscular fat which can be highlighted using dedicated techniques. To compensate the lack of signal due to the low amount of fat in muscle, a hybrid water-suppression sequence, which takes profit from both T1 and chemical shift contrasts, was proposed for being applied at high field (1). Otherwise an alternative frequency independent approach is based on the large difference between apparent diffusion coefficients of fat protons and water protons (2).
We previously proposed to image non fatty part of ECM on the basis of susceptibility differences between "hard" tissues and the myofiber bulk matrix (3). It was demonstrated that microstructure of muscle tissue can be investigated with high resolution T_2* imaging. This finding has been confirmed also in isolated rat heart (4), an important issue is now to understand the role of the different ECM compounds on susceptibility effects. The latter can be done by comparing MR images obtained at high field with stained sections of the same muscles obtained by histology.
Image analysis is a necessary step to characterize the spatial organization of the intramuscular connective tissue (IMCT) using morphological features. For this purpose pixels depicting the ECM in images of meat need to be extracted (or segmented) for quantifying quantitative parameters related to muscle organization i.e. perimysium distribution and distribution of the fascicle size delimited by the perimysium. The main drawback comes from the ECM volume, which is small compared to that of the myofibers.

Hence, even sensitive to ECM, images of muscle are characterized by a unique mode on their grey level histogram. Most of current segmentation approaches, especially the thresholding algorithms, fail in this context because of the requirement that two distinct modes (i.e. one for each tissue class which has to be segmented) should appear in the histogram, not too dissimilar in size and roughly normal in shape. The quantitative analysis of ECM distribution requires a specific technique. The paper sums up how a thresholding approach based on probabilistic reference maps allows an automatic identification of ECM in images (5).

2 METHODS

Biceps femoris (BF), *Infraspinatus* (IS), *Longissimus thoracis* (LT), and *Pectoralis profundus* (PP) muscles of Charolais cows, from 3 to 7 year old, were chosen for their different architectures, including fibre bundle size, EMC repartition and composition. Samples were aged 20-24 days *post mortem*.
The different bovine muscles were first imaged by MRI at room temperature (~20°C). Each sample was positioned so that the main muscle fiber direction was approximately parallel to the main magnetic field direction. The protocol consisted in acquiring high resolution axial T_2* weighted images (TR/TE = 5000/30 ms) at 4.7 T, leading to a small voxel volume of 0.1 x 0.1 x 0.5 mm^3. Afterwards the same samples were frozen and cut into section in the same axial plane. The composition of the connective tissue network was revealed on these sections by using red sirius for depicting the whole connective tissue and antibodies against more specific compounds of ECM i.e. collagen I and elastin. Despite the difference of intrinsic spatial resolution and field of view, the comparison of the two imaging modalities was possible because of the presence of external fiducial markers.

A thresholding method based on probabilistic reference maps was developed for characterizing ECM distribution on images of muscle exhibiting unimodal histograms. The process is based on a learning step. Probabilistic reference images were first derived by averaging the drawings obtained from a panel of 20 non-trained judges. For each reference, an optimal threshold could be defined which minimized the difference between the thresholded image and the reference card. By repeating the latter process on images of different animals and muscles, it was possible to define a multi-linear relationship between the optimal threshold and the histogram features of the images. This linear model gave rules which were used afterwards to calculate the optimal threshold for every image taken in the same standardized conditions (5).

3 RESULTS AND DISCUSSION

In spite of the geometrical deformations inherent to histology, the two imaging modalities reveal the same region of interest of muscle (see figure 1).

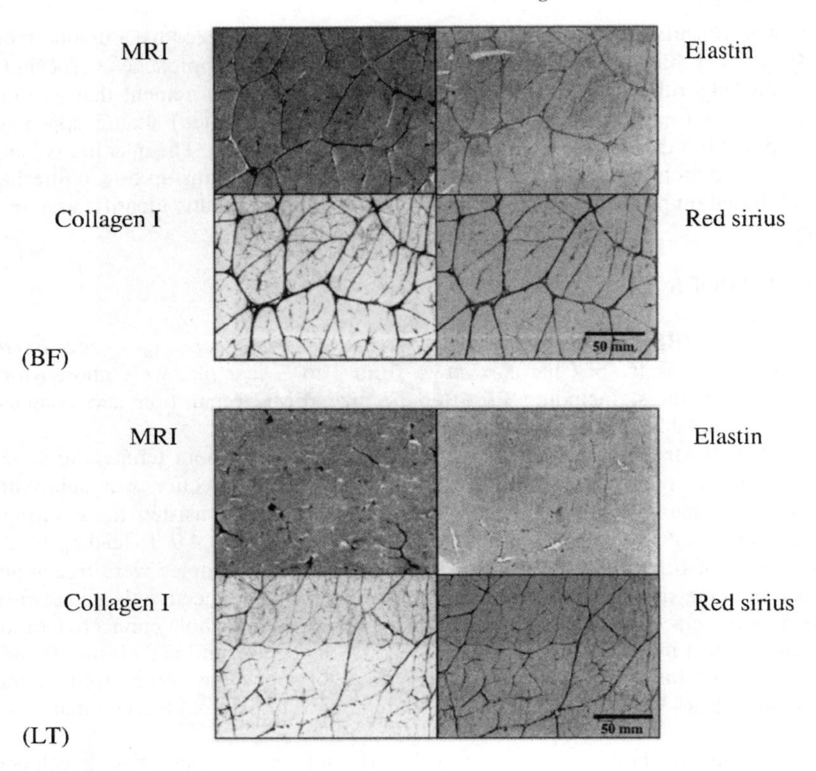

Figure 1 *Images obtained by MR and (immuno-)histology on Biceps Femoris and Longissimus thoracis allowing visual comparison.*

Map of collagen I, the bulk compound of ECM, and red sirius stain highlight typical repartition of connective tissue which is close to the one revealed by MRI only in BF muscle. In fact visual comparison underlines the dominant effect of elastin for provoking signal losses on susceptibility-weighted images due T_2* shortening. Elastin is mechanically strong and should influence the toughness of muscle for which it represents a significant part of the ECM, like in BF or *Semitendinosus* (6).

Because of suspected specificity of susceptibility-weighted images to elastin, complementary MR imaging techniques are needed in order to image other compounds of ECM. Ultra-short echo time MR techniques, such as single-point imaging (7) or acquisition schemes based on half-pulse excitation (8) are able to detect signal from protons in interaction with ECM characterized by short T_2 values. While promising for obtaining a detectable signal from ECM, the signal of myofibre protons remains dominant. Hence subsequent quantitative (T_2) mapping may be used for revealing different compounds of ECM. An alternative technique based on a double quantum filtering (9, 10) is able to detect signal coming from protons that experience anisotropic motion due to their interactions with ordered tissues of the ECM. The selectivity of this technique is promising for imaging ECM in muscle.

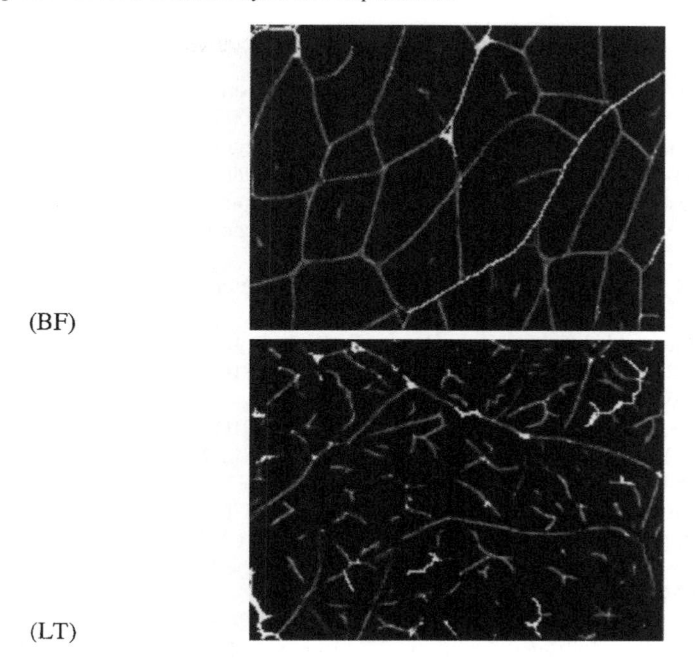

Figure 2 *Reference maps derived from MR images of bovine muscle depicting pixels belonging to ECM class. Maps were obtained by averaging drawings of 20 non-trained judges. The grey levels of the maps are then proportional to the consensus of manual segmentation between the judges.*

We have developed an automatic approach for global thresholding of unimodal images of ECM. Our segmentation method was objectively compared to the Rosin's unimodal thresholding algorithm (11) and validated by a histochemical quantification of IMCT collagen. Even if the building of a reference database from a set of manual drawings (see figure 2) is a tedious task, this step allows conducting more generic tasks; the optimization and/or the quantification of the precision of any segmentation method. This is the case when global thresholding is inefficient e.g. when the spatial inhomogeneities are significant or when more pixels have to be split in more than two classes. Besides, the same methodology has been used for other imaging modalities such as histology. A new set of reference maps and adapted thresholding rules, gave results as satisfactory as for MRI.

Once the optimal segmentation was obtained for both MRI and histology, morphometrical parameters characterizing the distribution of ECM were quantified and compared within several muscle types. The results of this image processing chain were applied to correlate the size of muscle fascicles with mechanical properties of meat (12). They indicate that, at a macroscopic scale, muscles with small fascicles delimited by a thick primary perimysium are tougher than muscles with large fascicles delimited by a thin primary perimysium.

References

(1) Laurent W., Bonny J.M. and Renou J.P., Magn. Reson. Imaging (2000) 12, pp 488-496
(2) Bonny J.M., Santé-Lhoutellier V., Renou J.P. (2005) Magnetic Resonance in Food Science – The Multivariate Challenge - Edited by Engelsen S.B., Belton P.S., Jakobsen H.J., pp 141-155 (proceedings of the 7th International Conference on Applications of Magnetic Resonance in Food Science held in Copenhague on the 13-15th September 2004)
(3) Bonny J.M., Laurent W., Labas R., Taylor R., Berge P., and Renou J.P., J. Sci. Food Agric. (2001) 81, pp 337-341
(4) Kohler S., Hiller K.H., Waller C., Jakob P.M., Bauer W.R., Haase A., Magn. Reson. Med. (2003) 49, pp 371-375
(5) Sifre-Maunier L., Taylor R.G., Berge P., Culioli J., and Bonny J.M., Image Vision Computing (2006) in press (available on line) doi:10.1016/j.imavis.2006.03.004
(6) Rowe R.W.D., Meat Sci (1986) 17, pp 293-312
(7) Gravina S. and Cory D.G., J. Magn. Reson. (1994) 104, pp 53-61
(8) Nielsen H.T., Gold G.E., Olcott E.W., Pauly J.M., Nishimura D.G., Magn. Reson. Med. (1999) 41, pp 591-599
(9) Neufeld A., Eliav U., Navon G., Magn. Reson. Med. (2003) 50, pp 229-234
(10) Mouaddab M., Foucat L., Bonny J.M. and Renou J.P., Proceedings of the Belgium-Luxemburg-Netherlands-French NMR conference of the GERM held in Blankenberge on the 19-22th March (2006) poster 52
(11) Rosin P., Pattern Recognition (2001) 34, pp 2083-2096
(12) Sifre L., Berge P., Engel E., Martin J.F., Bonny J.M., Listrat A., Taylor R. and Culioli J., J. Agric. Food Chem (2005) 53, pp 8390-8399

SEPARATION OF TWO DIMENSIONAL DIFFUSION AND RELAXATION TIME DISTRIBUTIONS FROM OIL/FAT AND MOISTURE IN FOOD

G.H. Sørland[1], F. Lundby[2] and Å. Ukkelberg[3]

[1]Anvendt Teknologi A/S, Hagebyv. 32, N-9404 Harstad, Norway, [2]MATFORSK, Osloveien 1, N-1430 Ås, Norway, [3]Dep. of Chemistry, University of Oslo, Norway

1 INTRODUCTION

One major drawback in relaxation time studies on food stuff using bench top low magnetic field NMR equipment is the overlap in relaxation times between the different constituents. Figure 1 illustrates the challenge one is facing when trying to interpret the T_2 distribution curve from a minced meat sample. Using the standard CPMG NMR sequence [1] and performing an ordinary inverse Laplace transform [2] on the data one ends up with four more or less resolved peaks (the distribution represented with a solid line in figure 1). From that distribution it is difficult to assign the peaks to fat and moisture, and resolving those components necessitates the use of pulsed magnetic field gradients (PFG) in PFG NMR experiments. Then one can make use of the significant difference in molecular mobility between fat and moisture and resolve the contributions to the distribution from the fat and moisture using a combined PFG-CPMG sequence [3, 4] .The result of such a procedure is also shown in figure 1 where the two other distributions result from the resolved contribution from moisture ('o') and from the resolved contribution from fat ('+').

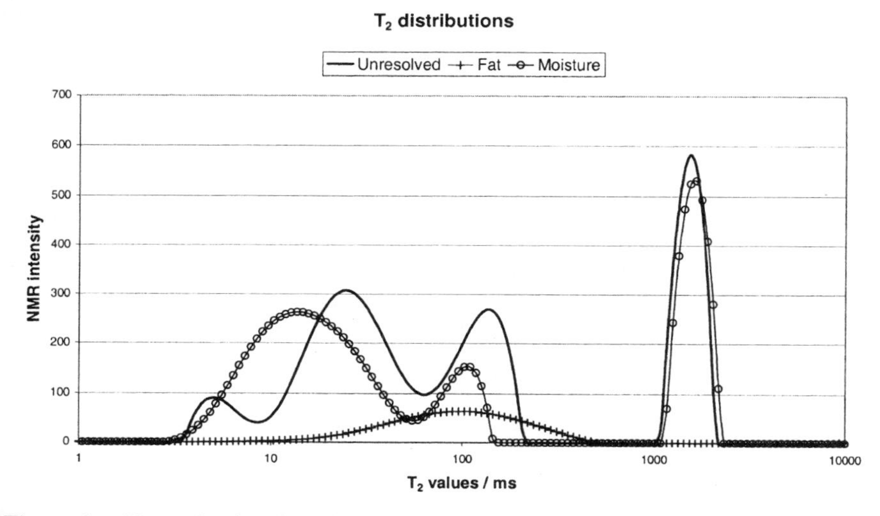

Figure 1 *Unresolved and resolved T_2 distributions from a baked minced meat sample.*

This method can be further developed in order to resolve the fat and moisture relaxation time curves at different gradient strengths in the PFG part of the sequences. Thus we may produce separated two dimensional diffusion – T_2 correlated distributions using the two dimensional inverse Laplace routine.

A drawback with this method is the possible loss of information during motional encoding/decoding in the PFG part of the sequence. During this time interval components with the shortest relaxation times (< 10 ms) may have attenuated to small amounts at the first acquisition point in the combined PFG-CPMG sequence. As one is acquiring the tails of the attenuation of these components it may result in a T_2 distribution in which the components at the shortest observation times have reduced in peak intensities. Thus we also present an improvement of the method presented by Seland et.al [3] where we take into account the attenuation during the PFG interval in the combined PFG-CPMG sequence. The major assumption leading to this improvement is that fat has small amounts of or none components with short T_2 values. This assumption is valid as long as the sample does not contain significant amounts of phospholipids and the fat is in a liquid state. With this method we are able to produce separated one dimensional T_2 distributions where the quantitative information is intact. That is, the distributions reflects total NMR signal from the sample.

2 THEORY

In work with relaxation time distributions acquired from low field NMR spectrometers it is quite evident that the fat and moisture signals are overlapping to various degrees, depending on the system studied. This makes interpretation difficult, and when trying to resolve the two components using a combined diffusion and relaxation time experiment (equation 1), the two dimensional inverse Laplace transform becomes unstable due to the lack of proper attenuation of the fat signal. This is due to limitations in the hardware of the bench top system where the maximum gradient strength is approximately 350 Gauss/cm. One could of course increase the duration of the gradient pulses to achieve proper attenuation of the fat signal as well. But the moisture signal with a typical short T^*_2 relaxation time during motional encoding/decoding would be significantly attenuated. Resolving this moisture component would therefore be difficult.

2.1 Separation of fat and moisture signal

For the reasons given above the method for resolving the fat and moisture signals prior to any post processing of the data [3, 4] have been implemented for food samples. Table 1 shows the combined Pulsed Field Gradient Spin Echo PFGSE-CPMG sequence (last figure), for which the attenuation from a multi-component system may be written.

$$I = \sum_{i=1}^{2} \sum_{k} \rho_{ik} e^{-\left(\frac{2(n\tau+2\tau')}{T_2^{ik}} + \gamma^2\delta^2(2g)^2 D_{ik}\left(\frac{3}{2}\tau' - \delta\frac{1}{6}\right)\right)} + \varepsilon \tag{1}$$

where $\boldsymbol{i}$ denotes either fat or moisture, $\boldsymbol{k}$ denotes the signal portion ρ_{ik} with a exponential decay characterised by the transverse relaxation time $\mathbf{T_2^{ik}}$, γ is the gyromagnetic ratio, D_{ik} is diffusion coefficient, g is the applied pulsed magnetic field strength, τ, τ' and δ are defined in table 1 (last figure), and ε represents the experimental noise. The numerical inversion of equation 1 is ill posed, meaning that several and very different solutions may fit the experimental data equally well [2,5]. An experimental approach, where the distribution from oil and water could be separated before the numerical inversion, should make the inversion less prone to error. This is done as follows: As the moisture normally exhibits a significantly higher self-diffusion than fat, the combined PFGSE-CPMG sequence is weighting the CPMG-decay to such a degree that the moisture gradually will contribute less and less to the CPMG-decay when applying an increased gradient strength in the PFGSE-CPMG sequence. When the gradient strength is sufficiently high, the signal contribution from the moisture becomes negligible, and the T_2-attenuation is originating from the fat only. If the diffusivities D_{1k} and D_{2k} differ significantly for all k, for example $D_2=50D_1$, it is possible to choose the applied magnetic field gradient g, and the time constant τ, such that signal from component 2 is suppressed to below a fraction of 0.01. Then there will still be a fraction of 0.5 ($D_2=50D_1$) of component 1, for which the relaxation curve is recorded. As the mobility of fat is significantly lower than the mobility of moisture, it is possible to subtract a scaled version of the CPMG decay from fat only, in the combined PFGSE-CPMG sequence with the pulsed field gradients set to zero. The resulting T_2 attenuation will then originate from the water phase only. Following the procedure for separation of multi-component attenuation [3] we are able to measure the T_2 contributions from oil and water separately at any of the applied gradient strengths. As a consequence we may perform two dimensional inverse Laplace transform (2D-ILT) [5] on fat and moisture separately, which leads to more correct transforms than transforming the unresolved contribution. However, the methods suffers from the fact that in the presence of moisture signal with very short T_2 values, the 2D-ILT is not quantitatively correct. The moisture signal at shorter T_2 values will be underestimated. In the next section we present a new method that produce one dimensional T_2 distributions where all components are quantitatively correct, i.e. equally weighted.

2.2 Resolving the contribution of fat and moisture from a CPMG attenuation

In order to resolve the contributions from fat and moisture it is vital to know the exact quantities of the fat and moisture within the sample. This is done using the Oneshot method [6]. Furthermore we must model the CPMG attenuation from fat by making use of the attenuation from fat in the combined PFGSE-CPMG with the moisture signal suppressed. This model is found using Kaufmann's algorithm [7] which fits a sum of exponentials to a curve. The idea of the algorithm is to express the sum of exponentials as a linear combination of powers of the independent variable and successive integrals of the sum. In the present implementation, the number of components (NCO) is estimated using the technique known as cross validation. A series of subsets of the measurements are successively left out, and inverse Laplace fits on the remaining data are done with increasing values of NCO. These reduced models are used to predict the values in the subset which is left out. This is done until all the measurements have been left out and predicted. The NCO giving the smallest error of prediction is then chosen as the optimal NCO. Finally, a fit is made using the optimal NCO and the complete set of data. This algorithm is compact, fast and non-iterative, and has been tested on both synthetic and measured data, and has so far shown very satisfactory performance.

Figure 2 shows the attenuation of fat only and the fitted model found using Kaufmann's algorithm. This model is then scaled according the initial fat signal found from the Oneshot method [6] and thereafter subtracted from the ordinary CPMG sequence. The resulting attenuation is then from moisture only. If the inter echo spacing in the CPMG sequence is short, the shortest T_2 component may arise from the protein signal in the protein/moisture interface, or if there are significant amounts of phospholipids, they will also contribute to the shortest component.

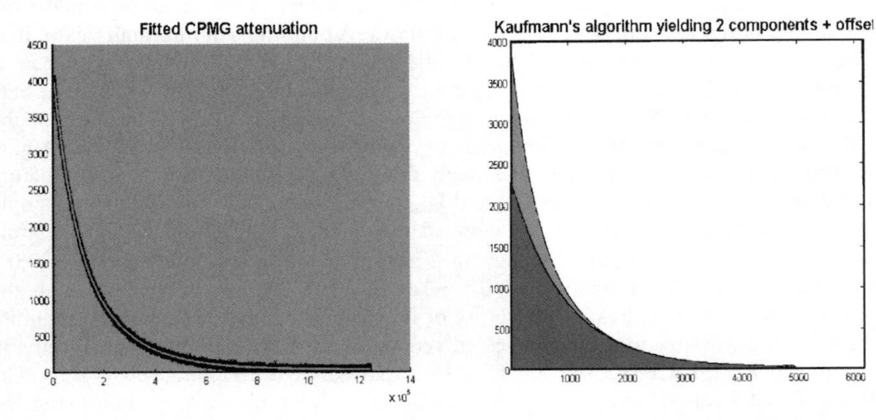

Figure 2 *Kaufmann's algorithm applied on a combined PFGSE-CPMG attenuation where moisture has been suppressed using strong magnetic pulsed field gradients in the PFGSE.*

3 EXPERIMENTAL

To resolve the contribution from moisture and fat in food stuff, we analyzed samples containing salmon and minced meat. The minced meat sample was measured as a raw sample and after it had been exposed to 105°C for 1 hour. The instruments used were a Maran Ultra 23 MHz capable of delivering 350 Gauss/cm of applied gradient strength, and a Maran DRX 12 MHz capable of delivering 240 Gauss/cm of applied gradient strength. Samples having a diameter of 18 mm were measured using the 23 MHz and at an operating temperature of 40°C while samples having a diameter of 38 mm were measured using the 12 MHz at an operating temperature of 35°C. The pulse sequences used are listed in table 1 with their corresponding attenuations.

4 RESULTS AND DISCUSSION

As indicated in the theoretical section, the successful resolution of fat and moisture is dependent on a significant difference in molecular mobility. On a substantial number of samples containing meat, fish and diary products we have found that the diffusion coefficient of liquid fat is approximately 2.3 10^{-11} m^2/s while it is 3.0 10^{-9} m^2/s for moisture at a measuring temperature of 40°C. Figure 3 shows a typical two component fit of such a data set, where the mobility of the fat is two orders of magnitude lower than the mobility of brine. Following the procedure for resolving the fat and moisture signal contributions we apply the combined PFGSE-CPMG sequence to achieve separate Diffusion-T_2 correlated data sets. On these data sets we may then apply commercially available 2 Dimensional Inverse Laplace Transform (2D-ILT) [5]. Such a 2D-ILT analysis is shown in figure 4 on a sample of salmon tissue, and for comparison we also show the 2D-ILT applied on a data set where fat and moisture are not resolved prior to the numerical analysis.

Table 1 *The NMR pulse sequences used for resolving the contribution from fat/oil and moisture.*

NMR sequences	Corresponding attenuations
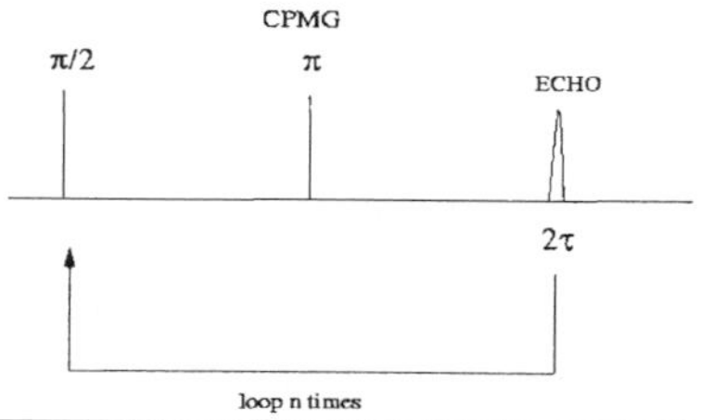 CPMG $\pi/2$ π ECHO 2τ loop n times	The ordinary CPMG $I = I_0\,e^{-n\bullet[\frac{2\tau}{T_2}-\frac{2\tau^3}{3}\gamma^2 G_i^2 D]}$
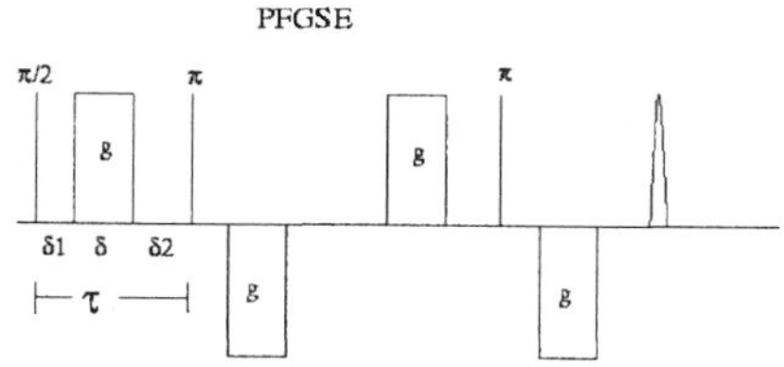 PFGSE $\pi/2$ π π g g g g $\delta1$ δ $\delta2$ τ	The Pulsed Field Gradient Spin Echo $I = I_0 e^{-(\frac{4\tau}{T_2^{ik}}+\gamma^2\delta^2(2g)^2 D_{ik}(\frac{3}{2}\tau-\delta\frac{1}{6}))}$
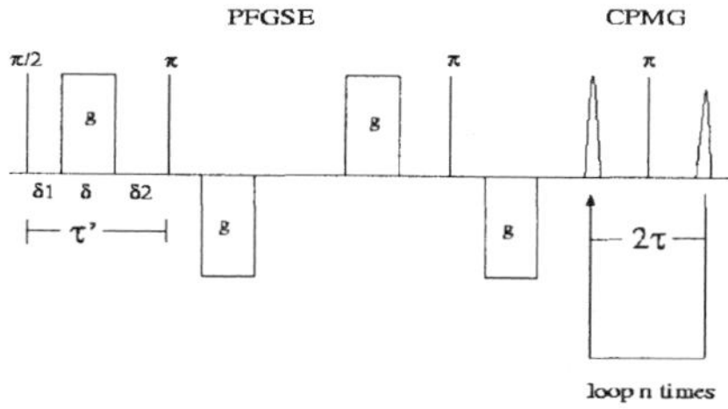 PFGSE CPMG $\pi/2$ π π π g g g g $\delta1$ δ $\delta2$ τ' 2τ loop n times	The Combined Pulsed Field Gradient Spin Echo – CPMG $I = I_0 e^{-(\frac{4\tau'}{T_2}+\gamma^2\delta^2(2g)^2 D(\frac{3}{2}\tau-\delta\frac{1}{6}))}$ $e^{-n(\frac{2\tau}{T_2}+\frac{2\tau^3}{3}\gamma^2 G_i^2 D)}$

As seen in figure 4 the 2D-ILT resolves moisture into two components with respect to transverse relaxation times, but with approximately the same diffusion coefficient, 3.2 10^{-9} m^2/s (at 40°C). Without separation of fat and moisture prior to analysis, the fat component does not result in a stable solution and its diffusion coefficient varies almost two orders of magnitude. With separation of fat and moisture prior to 2D-ILT analysis, the fat component also yields a stable solution and a diffusion coefficient of 2.5 10^{-11} m^2/s (at 40°C). The reason for the failure of the 2D-ILT applied on the total signal is most likely due to the lack of sufficient attenuation of the fat signal. In the presence of a large and fast attenuating moisture signal the 2D-ILT routine is not able to discriminate between a constant offset signal and a signal which is attenuating much more slowly. Then the position of the 2D distributions of the fat becomes dependent on the input parameters to the 2D-ILT routine.

One could of course argue that with a proper attenuation of the lipid signal, the 2D-ILT routine would also be able to fit the lipid signal quite accurately and return a stable solution. Such an approach would either require much stronger magnetic field gradients or a longer duration of the gradient pulses in the PFGSE interval. The option of much

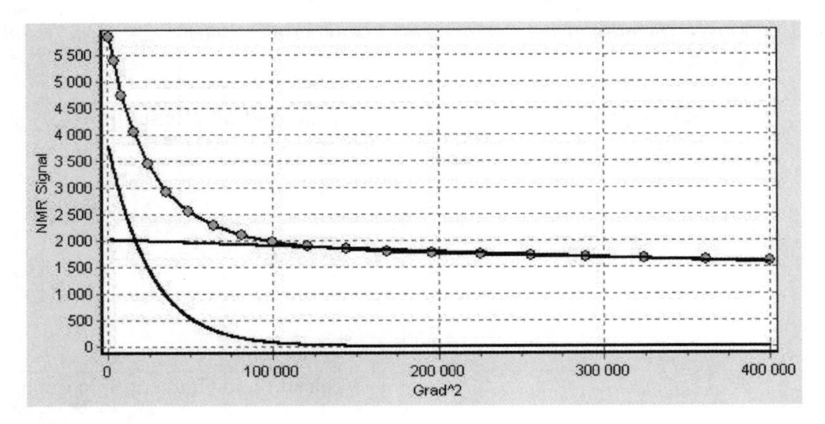

Figure 3 *PFGSE experiment on minced meat. The slopes indicate a fast diffusing component (fast attenuating signal as a function of the square of the applied gradient strength (Grad^2)) and a slow diffusing component (slow attenuating signal as a function of Grad^2)*

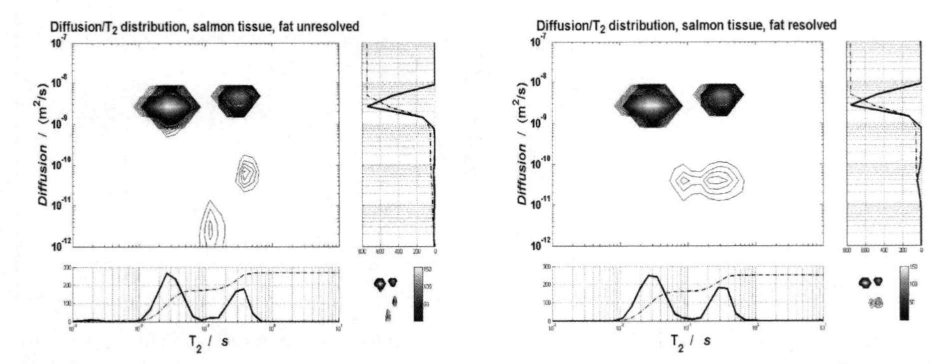

Figure 4 *The result of 2D-ILT of salmon tissue of low fat content; without (left) and with (right) separation of fat and moisture prior to analysis*

stronger magnetic field gradients would require a new gradient coil configuration with smaller coil dimensions to achieve 1000 Gauss/cm and thus reduced sample size (~ 5-10 mm in diameter) to get the normalized lipid signal(I/Io) attenuated down to below -1 on the logarithmic scale. Applying the second option, longer gradient pulse duration, would result in a first measuring point at more than 30 ms in the CPMG part of the sequence. At that time much of the NMR signal arising from the moisture with the shortest T_2-values has already attenuated, thus resulting in a diffusion-T_2 correlated distribution that is less representative for the sample investigated.

In figures 5-6 we show the diffusion - T_2 correlated distributions from a minced meat sample before and after being exposed to 105°C for an hour. The fat and moisture signals have been resolved prior to 2D-ILT. Using the Oneshot method for quantification [6], we found the fat content to be 14 % while we found the moisture content to be 70 %.

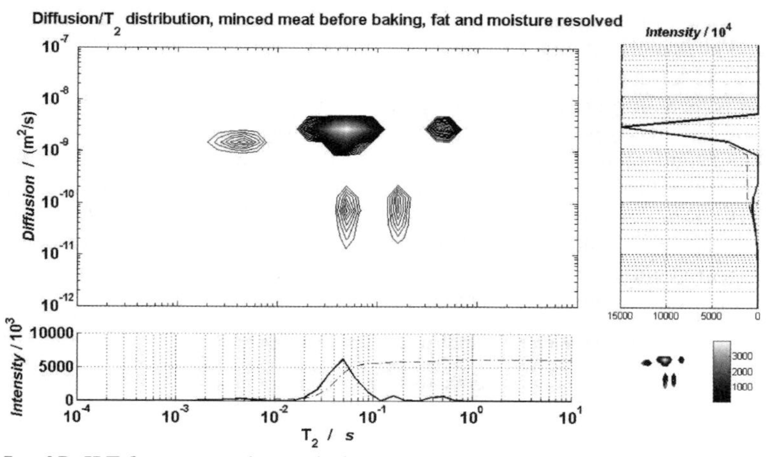

Figure 5 *2D-ILT from minced meat before baking at 105°C.*

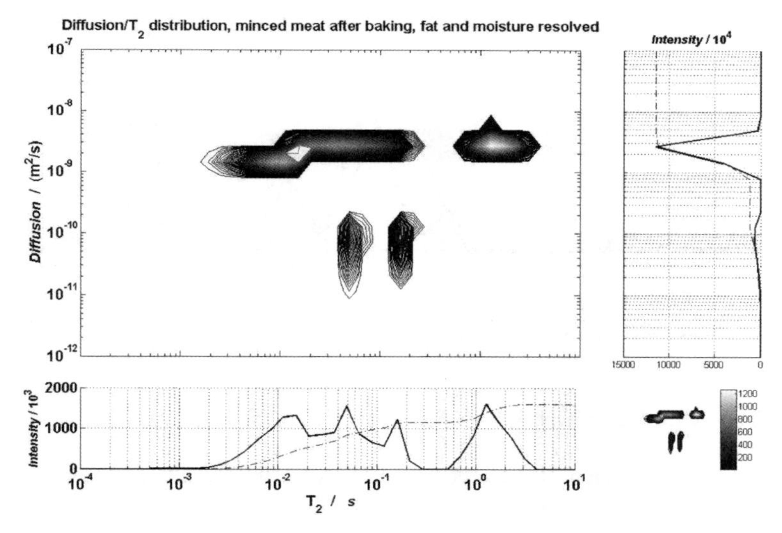

Figure 6 *2D-ILT from minced meat after baking at 105°C.*

As already shown in figure 4, applying the same 2D-ILT routine on an unresolved data set does not return distributions where the fat and moisture signals are resolved. For both unresolved 2D-ILT distributions, the majority of the fat component was found at the lowest diffusion coefficients possible, i.e. at the edge of the distribution plot. If the field of view was changed to take into account even lower diffusivites, some of the fat component followed the edge while another portion was located by the 2D-ILT routine to diffusion regions where it overlapped with moisture signal. Thus the interpretation of such distributions becomes heavily dependent on the input parameters to the 2D-ILT routine, such as the degree of smoothing and field of view. On the other hand we may deduce from the resolved distributions in figures 5-6 that after baking of the meat some of the moisture leaks out (longest T_2), another portion found at shortest T_2 values and lower mobility than the bulk moisture has increased significantly, and a reduced intermediate portion remain unchanged in T_2 and mobility.

Due to a finite duration of approximately 10 ms of the PFGSE part of the PFGSE-CPMG sequence, information from components with T_2 values less than 10 ms is reduced or has vanished. Thus we have applied the method for modelling the CPMG attenuation from fat and subtracting it from the ordinary CPMG attenuation. In figure 7 the resulting 1 dimensional moisture distributions after baking the minced meat are displayed. Using the CPMG data we find another component at 0.2 ms which is not resolved using the distribution achieved with the PFGSE-CPMG sequence. The origin of this component is not clear as it could arise from the tail of protein signal in the moisture-protein interface, phospholipids, or from moisture bound to the solid matrix.

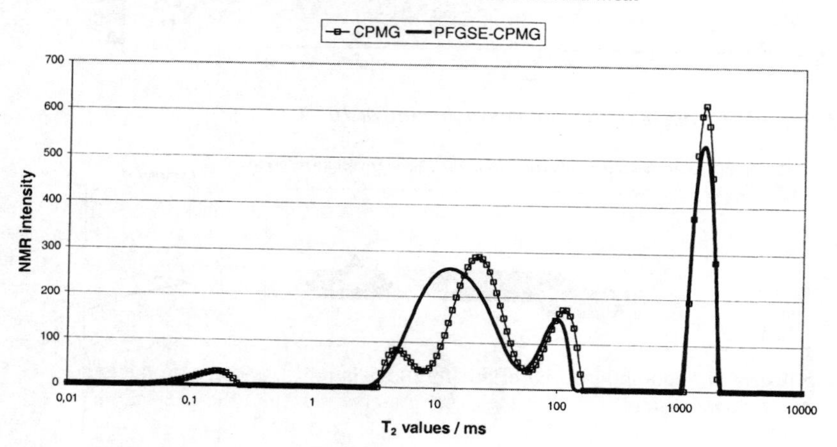

Figure 7 *Resolved T_2 distributions from moisture in baked minced meat.*

5 CONCLUSION

When analysing diffusion T_2 correlated data sets, the result becomes stable and reproducible when separating the moisture and fat signal prior to applying 2D-ILT.

References

1 Meiboom S., Gill D., *Rev. Sci. Instrum.* 1958; 29, 688
2 Provencher S.W: *Comput. Phys. Commun.* 1982; 27, 213
3 Seland J.G., Sørland G.H., Antonsen H.W., and Krane J., *Applied Magnetic Resonance* 2004;24, 41-53
4 Sørland G.H., Anthonsen H.W., Seland J.G., Antonsen F., Widerøe H.C., and Krane J., *Applied Magnetic Resonance* 2004;26, 417-425
5 Song Y.Q., Venkataramanan L., Hürlimann M.D., Flaum M., Frulla P. and Straley C., *Journal of Magnetic Resonance* 2002;154, 261-268
6 Sørland G.H., Larsen P.M, Lundby F. Rudi A., Guiheneuf T., *Meat Science* 2004;66, 543-550
7 Kaufmann B., http://arxiv.org/abs/physics/0305019 , 2003

DAIRY PRODUCT AUTHENTICATION BY ^{1}H NMR SPECTROSCOPY IN COMBINATION WITH DIFFERENT CHEMOMETRIC TOOLS

M. Cuny[1], E. Vigneau[2], M. Lees[1] and D.N. Rutledge[3]

[1]Eurofins Scientific Analytics, rue Pierre Adolphe Bobierre, BP 42301, 44323 Nantes cedex 3, France
[2]ENITIAA/INRA, Sensometrics and Chemometrics Laboratory, BP 82 225, 44 322 Nantes cedex 3, France
[3]Institut National Agronomique Paris-Grignon, UMR INRA - INA P-G N°214, 75341 Paris cedex 5, France

1 INTRODUCTION

As an important dietary source of calcium, protein and vitamins A, B12, and D, the consumption of milk and dairy products is on an upward trend. This increase is mirrored by the growing number of novel fresh dairy products (fat-free, full-fat, set, stirred, drinkable, probiotic) available on the supermarket shelves, very often containing varying proportions of fruit, flavours or both to boost sensory appeal. The authentication of such products generally requires analysis of a large number of parameters linked to the components in the product. Therefore the availability of a rapid analytical method for this purpose would be a great help in ensuring compliance with labelling legislation. As dairy products are rich in fat, few methods are suitable for broad screening techniques, and the measurement of fruit content itself is a notoriously difficult task.

NMR spectroscopy is emerging as a popular technique for food analysis as it can be used for both quantitative compositional analysis and for rapid throughput screening[1]. Combined with chemometric techniques it can be a useful and rapid method to assess food authenticity and it has already been applied to fruit juices and to olive oil in particular to detect various types of adulteration.

The aim of this study was to assess the potential of ^{1}H NMR spectroscopy in analysing fruit-containing dairy products. The measured NMR response is assumed to be a function of the relative proportions of the components in the mixture. The work presented here focuses on different chemometric tools used to extract relevant information from the ^{1}H NMR spectrum. The results of these different methods are assessed in terms of their ability to discriminate between types of yoghurt based on their fruit content.

2 METHODS

2.1 Sample Preparation

65 strawberry yoghurts (16 flavoured yoghurts, 15 stirred fruit pulp yoghurts, 34 fruit yoghurts) were collected from local supermarkets and frozen prior to preparation (Table 1). The samples were assigned to a percentage group (0%; <10%; >10%) according to the fruit content indicated on the label. Sample preparation consisted of 25 min of centrifugation at 4500 rpm, after which 750 μL of supernatant was collected. To this was added 250 μL of

D_2O containing 0.75 % of TSP (sodium 3-(trimethylsilyl)propionate-2,2,3,3-d4). No additional treatment was necessary. The D_2O was used as a source of the field frequency lock signal, and TSP for internal referencing of ^1H chemical shifts.

Table 1 *List of samples and associated fruit content group*

Sample Type	% Group	Number of samples
Fruit yoghurt	<10%	16
Fruit yoghurt	>10%	18
Total:		**34**
Stirred fruit pulp yoghurt	<10%	14
Stirred fruit pulp yoghurt	>10%	1
Total:		**15**
Flavoured yoghurt	0%	**16**

2.2 Data Acquisition

^1H NMR spectra were run on a Bruker Advance DPX-400 spectrometer using the noesypr1d sequence with water peak suppression during relaxation delay and mixing time. For each spectrum 520 transients were accumulated with 32 K data points. The acquisition time was about 3.5 sec and the spectral width was 12 ppm. Each experiment was carried out in 45 min.

2.3 Data Pre-treatment

The phase was corrected by hand using Bruker Topspin software. All other data treatment was carried out using Matlab version 7. These consisted of base-line correction, warping using the Correlation Optimised Warping function with linear interpolation (COW[2,3]) and data reduction. This data reduction was obtained by using the mean of 11 adjacent points to concentrate the information contained in the ^1H NMR spectrum. Finally a logarithmic transformation was applied to the data to reduce the difference in scale between the different parts of the spectrum. In this way the initial spectrum of 32 K points was reduced to 2241 variables, the transformation is shown in Figure 1.

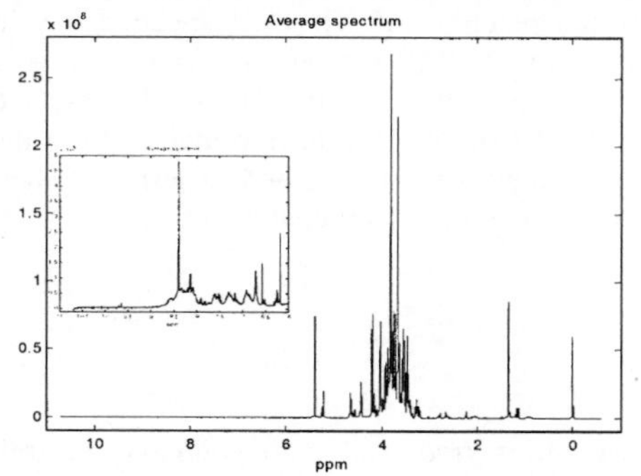

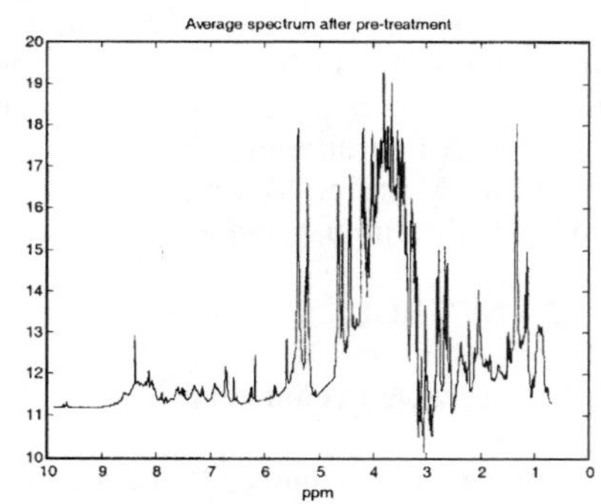

Figure 1 *(a) Averaged ^1H NMR spectrum of all samples; (b) Averaged data set after pre-treatment with logarithmic transformation*

2.4 Chemometric treatment of data

2.4.1 *Variable Selection.* Three different approaches were investigated to select NMR variables in order to predict yoghurt type or fruit percentage level.

Selection of variables with high variance (Variance method) – Since the variance of a variables is a measure of the average squared distance between the individual values of the variable for the set of data points and their mean value, this criterion can be used to identify the most discriminant variables in the data set. Indeed the latter are expected to vary more than noisy or non-informative variables. A variance threshold, above which variables were kept in the data set, was defined by the ability of the resulting data to predict membership of the samples to the correct group (both for yoghurt type or for fruit percentage) using the cross-validation technique.

Selection of clustered variables (CLV[4,5] method) – A cluster analysis around latent variables is applied in order to identify groups among the spectral variables, linked either by their covariance or their correlation. Organising multivariate data into a small number of clusters, each represented by a latent component, makes it possible to reduce dimensionality. The CLV method involves two stages, namely a hierarchical clustering analysis followed by a partitioning algorithm. Partitioning is determined by the value of a quality criterion, in this case T, which is the sum of the first eigenvalues of the data matrices of all the clusters. Then the component most likely to discriminate between samples in terms of yoghurt type or fruit percentage is determined by Analysis of Variance (ANOVA).

Selection of variables defining zones of continuous variables (Evolving Window Zone Selection - EWZS method) – This approach focuses on detecting different continuous parts of the spectrum that are shown to be most discriminant either for yoghurt type or for fruit percentage. This function is used to do a singular value decomposition – based PCA and then both a cross-validation PLS on all the scores and simple linear regressions between each vector of scores and the variable to be predicted (yoghurt type or fruit percentage). The function plots maps of the Root Mean Standard Error of Cross Validation (RMSECV) values calculated by cross-validation PLS and the correlation coefficients (R^2) calculated by linear regressions for each area of the spectrum. EWZS uses windows of varying size, from small user-defined minimum up to a user-defined maximum (which may be the width of the data set). The first set of windows starts with the first variable. Once the window has increased to the user-defined limit, the starting point of the new set of windows is incremented by a user-defined step. In that way visual examination of these maps is used to identify those zones showing a minimum RMSECV value and a maximum R^2 value which are then selected as informative continuous zones of variables.

2.4.2 *Comparison of Variable Selection Methods.* Factorial Discriminant Analysis (FDA) was then carried out on the principal components of the parts of the spectrum selecting using the methods described above. To compare the discriminant power of the three approaches the number of correctly classified samples was calculated using a leave-one-out cross-validation method.

3 RESULTS AND DISCUSSION

3.1 Discrimination of yoghurt types and fruit percentage groups after selection based on high variance criterion

The variance of the variables is shown in Figure 2. The highest variability is found in the region from 0 to 5.5 ppm of the spectrum. This corresponds mainly to sugar and organic acid content. The number of correctly classified samples using the leave-one-out cross-validation technique was calculated for a variance ranging from 0.01 to 0.1 and the classification carried out for yoghurt types and fruit percentages. The best result of classification was obtained for a threshold of 0.045 for the assignment of yoghurt type. For the fruit percentages the number of correctly classified sample is relatively stable between the threshold values ranging from 0.035 to 0.050, and therefore the value of 0.045 is chosen.

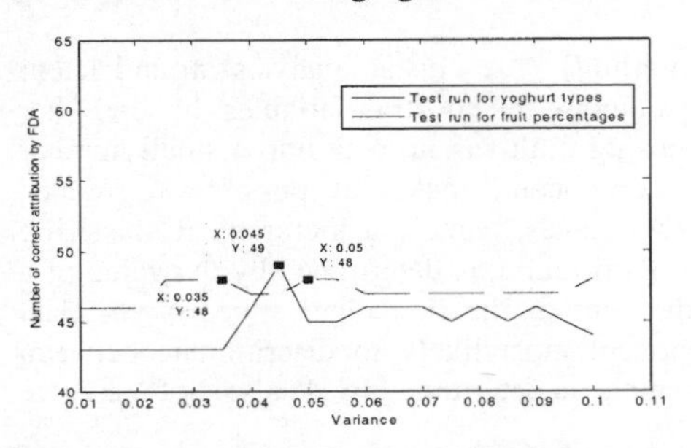

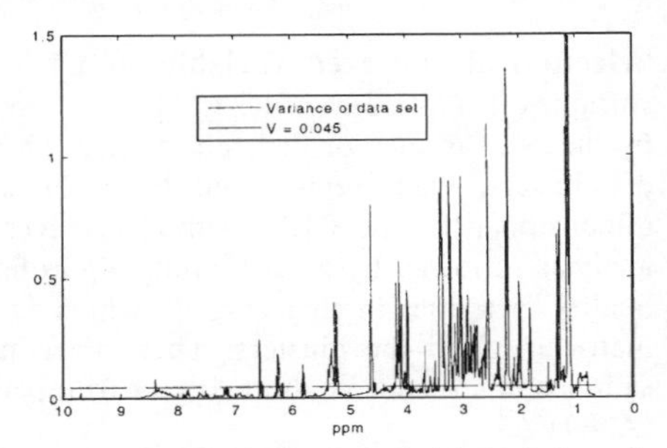

Figure 2 *a) Results of the selection algorithm to define the variance threshold; b) Selected variance threshold on variance data set.*

The results for a variance threshold of 0.045 are presented here (Table 2). This led to the selection of 638 variables (Figure 5). The first three principal components of the reduced data set were submitted to FDA to predict either the yoghurt type or the fruit percentage. In the case of yoghurt type, discrimination into two groups (flavoured vs. others) was also tested. The calculation time of this method was very rapid, about 3 minutes.

Table 2 *Summary of results for Variance method*

Number of variables selected	638	28.5%
FDA on yoghurt types	F1 80 %	F2 20 %
N° of samples correctly-classified		
For 3 groups	49	75 %
For 2 groups	62	95 %
FDA on fruit percentages	F1 79 %	F2 21 %
N° of samples correctly-classified	47	72 %
Calculation time:	rapid	

3.2 Discrimination of yoghurt types and fruit content with selection based on CLV method

CLV analysis has two options for grouping variables, either by covariance or correlation. Both analyses were carried out but only the first case giving the best results, is presented here.

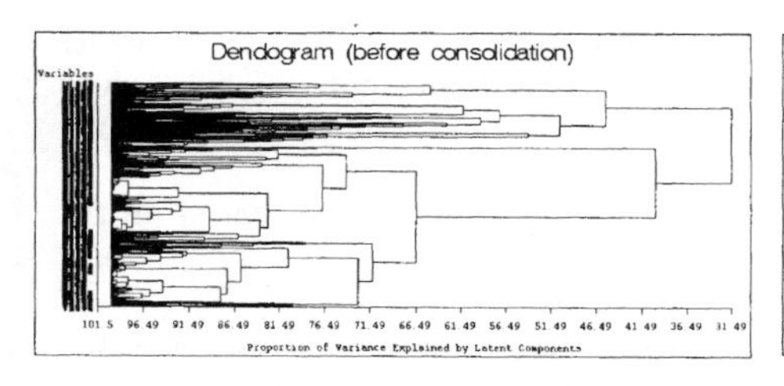

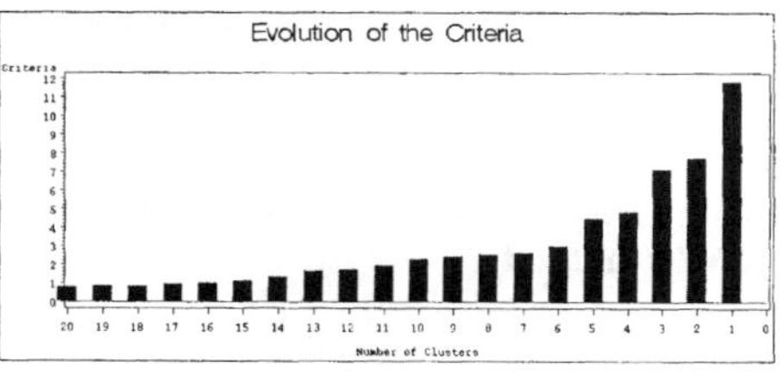

Figure 3 *a) Dendrogram of variables, b) Evolution of the criterion ΔT with the number of clusters*

The calculation time of this part of the analysis is rather long, but can be decreased by using other algorithm options.

The graph of ΔT (Figure 2b) observed during the hierarchical clustering showed that when passing from a partitioning of 4 to 3 groups the T criterion significantly increased. An ANOVA on the latent variables associated to the four groups of the retained partition was used to identify the significant components for explaining yoghurt type.

Table 3 *ANOVA results for yoghurt type on the four latent obtained by CLV*

Source	Sum of Squares	d.f.	Mean Squares	F	Prob>F
Component 1	0.014	2	0.007	0.44	0.649
Component 2	0.073	2	0.036	2.44	0.0955
Component 3	0.371	2	0.185	18.28	5.74E-07
Component 4	0.840	2	0.420	162.93	0

The latent components of the 3rd and 4th groups were significant at α= 5 %. Therefore the variables of both groups were selected. Next, variables were linked manually to reshape some of the spectral areas. This was done for both groups. Their discriminatory power was tested both separately and jointly. The best results shown here were obtained with the 275 variables from the 4th group (Figure 5).

An FDA was carried out based on yoghurt types and fruit percentages. The predictive power was assessed by the number of correctly classified samples in a leave-one-out cross-validation, results in Table 4.

Table 4 *Summary of results for CLV method*

Number of variables selected	275	12 %
FDA on yoghurt types	F1 91 %	F2 9 %
N° of samples correctly classified		
For 3 groups	44	68 %
for 2 groups	64	98 %
FDA on fruit percentages	F1 79 %	F2 21 %
N° of samples correctly classified	54	83 %
Calculation time:	long	

With the 275 selected variables representing 12 % of the initial variables, the discrimination of flavoured *versus* pulp fruit stirred and fruit yoghurt is 98 %.

When the variables selected from the third component were added to the variables selected from the fourth component the results were not as good but similar to those which could be obtained by using the 2241 initial variables.

3.3 Discrimination of yoghurt types and fruit content with selection based on the EWZS method

Applying the EWZS function to the data set gave with a relatively short calculation time, the maps shown in Figure 4. The minimum window size was set to 11 variables, the maximum window size to 250 while the increment between each series of evolving windows was a step of 20 variables. Visual examination of the R^2 map shows that variables ranging from 1900 to 2141 along with the zone from 520 to 1000 are the most discriminant for yoghurt type. The minimum RMSECV plot shows that the variables 1900 to 2141 along with the zones from 1220 to 1340, and from 1620 to 1760 are the most discriminant for yoghurt type.

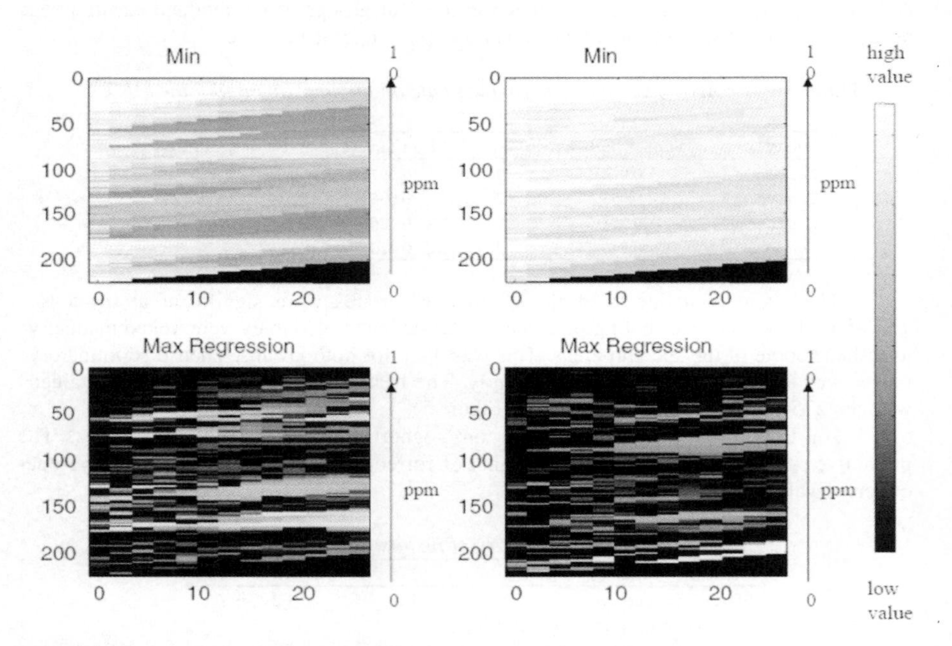

Figure 4 *Maps of minimum RMSECV and maximum correlation coefficients (R^2) for varying size zones up to 250 variables wide and initially starting at the first point: a) prediction of yoghurt type, b) prediction of fruit percentage*

The zones selected in this way are shown in Figure 5 for both yoghurt type and percentage fruit groups. A summary of results for both groups is given in Table 5.

Table 5 *Summary of results for EWZS method*

Number of variables selected	1105	49 %
FDA on yoghurt types	F1 92 %	F2 8 %
N° of samples correctly-classified		
for 3 groups	46	71 %
for 2 groups	64	98 %
Number of variables selected	722	32 %
FDA on fruit percentages	F1 95 %	F2 5 %
N° of sample correctly-classified	51	78 %
Calculation time:	intermediate	

4 CONCLUSION

The three sets of selected variables gave similar results for both predictions, with more than 70 % correctly classified samples. If only two types of yoghurts (flavoured and fruit yoghurts) are taken into account, the best results were obtained using the more complex algorithms for the selection of variables, with 98 % of the samples correctly classified.

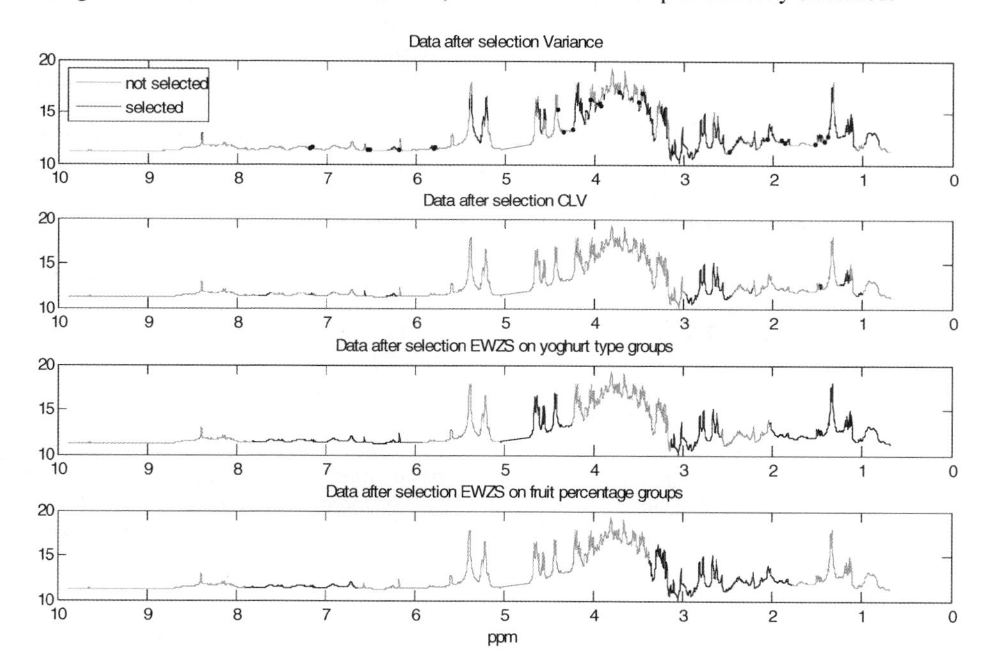

Figure 5 *Selected variables on the averaged spectrum and results of discriminations by applying the different selection methods*

All of the approaches investigated enabled a reduction in data set size, ranging from 51 % for the EWZS method to 88 % for the CLV method. Since the EWZS function selects continuous parts of signals, if two relatively close peaks give good predictions then the zone containing both will also be picked as a good predictor. This means that the variables in between will also be included. This is not the case with the CLV and the variance methods, which therefore give smaller data sets. In addition, if the information being used to discriminate samples is not especially close, the result of the EWZS function may not be as good as when only the significant marker peak is retained.

Two main zones containing information that discriminate yoghurt types were found by all three methods: chemical shifts between 2.5 and 3 ppm corresponding to major organic acids, and the zone between 1.1 and 1.3 ppm. To a lesser extent, some parts of the aromatic signals were selected in the low field region. None of the major sugar components were selected by the CLV method for discrimination (area between 3 and 6 ppm).

The two methods, CLV and EWZS, which involve some multi-dimensional optimisation criteria for selection, provide more information to interpret the results. The major difference between these two methods is in the calculation times – the EWZS methods being significantly much faster. However another way to do CLV is to perform, first, a partitioning algorithm with many groups (more than the number of the individuals), but far less than the number of variables. Then the latent components of these groups are submitted to the classic CLV algorithm. In the first stage, an initial partition is chosen at random and the quality criterion to be optimised is the same criterion as in classical CLV. This method is almost as fast as the variance method although the results will depend on the initial partition.

Future work will be on testing the feasibility of using the Variance method and/or the EWZS method as input to the slower but more selective CLV method as such a prior reduction in the size of the data set should significantly reduce the calculation time.

The samples used in this study are commercially available products with no guarantee of their authenticity. This work will be further backed up by a study of authentic samples of defined fruit content. These preliminary results, however, show that ^{1}H NMR with suitable data treatment is a promising approach to the quantification of fruit content in yoghurt and other complex matrices.

References

(1) Le Gall, G.; Colquhoun, I. J. In *Food authenticity and traceability*, pp 131-155.

(2) Tomasi, G.; van den Berg, F.; Anderson, C. *J. Chemometrics* **2004**, *18*, 231-241.

(3) Nielsen, N. P. V.; Carstensen, J. M.; Smedsgaard, J. *J. Chromatography* **1998**, *805*, 17–35.

(4) Vigneau, E.; Sahmer, K.; Qannari, E. M.; Bertrand, D. *J. Chemometrics* **2005**, *19*, 122-128.

(5) Vigneau, E.; Qannari, E. M. *Communications in statistics* **2003**, *32*, 1131-1150.

A TERNARY FULL-RANK EXPERIMENTAL DESIGN AS VIEWED BY CHEMOMETRICS AND NMR SPECTROSCOPY

H. Winning and S.B. Engelsen

Quality & Technology, Department of Food Science, The Royal Veterinary and Agricultural University, Rolighedsvej 30, 1958 Frederiksberg C, Denmark

1 INTRODUCTION

Data is not the same as information. A most critical manoeuvre in chemical as well as physical analysis is to master the analytical methods and the integration of the data analysis. One-to-one relationships between a cause and a measurement have hitherto worked well in science, but are now a hindrance to the study of more complex systems such as those within nutrigenomics which are characterised by hidden many-to-many relationships. To find these relationships unsupervised data exploration and data mining techniques are required.

NMR is a unique and versatile spectroscopic method capable of measuring samples in the solid, liquid and gas phases. NMR spectroscopy is able to measure small molecules as well as large biomacromolecules and requires a minimum of sample preparation. There exists no other spectroscopic method that contains so detailed structural information including conformational changes and dynamic chemical exchanges. The chemical shifts are characteristic to the molecular conformations and the signal intensity is proportional to the concentration for which reason NMR spectroscopy is quantitative and obeys Beer's law. However, a serious challenge in NMR spectroscopy lies between the technical capacity to generate data and the human capacity to interpret and integrate these data. In a split second ^{1}H NMR can generate 65.000 variables which hold information about all the protons in a sample. This amount of data is difficult, if not impossible, to interpret for complex systems. Expert knowledge about the system as well as about NMR is necessary and the human interpretation (assignment) of the spectra is often very time-consuming.

Statistical evaluation of large data sets generated by experimental studies is of key importance in order to obtain meaningful and interpretable results. Chemometrics uses mathematics and statistics to compute and extract chemical information from large two-dimensional multivariate data sets. By means of chemometric data analysis software program it is possible to simplify the immense number of variables in spectral data into a few significant factors (latent variables) and facilitate a rapid and interactive overview of data structures by the graphic computer interface. Development of chemometric data models requires a minimum of assumptions and the results and causal relationships may be visualised by intuitive interpretable illustrations. The detection of patterns in so-called score plots can be used to determine important effects for the discrimination between different groups of samples and for investigating trends in data. In this paper we provide a simple tutorial in exploratory analysis of NMR spectra, explaining the most fundamental, simple and common chemometric methods and principles. For this purpose we have chosen a designed tri-axial model design with three simple linear water soluble alcohols containing different amounts of hydrocarbons: methylene and methyl groups. Using this design we can explore subtle differences in the methylene peak – a simplified simulation of one of the major metabolomic applications of NMR, namely lipoprotein profiling of blood. Quantitative analysis of lipoprotein major fraction, LDL, VLDL and HDL is of great interest for medical purposes, for instance in coronary heart diseases, diet management and cancer. Using chemometrics to analyse NMR spectroscopy can provide qualitative and quantitative information on the methylene pattern of the different lipoprotein fractions[1].

2 METHOD AND RESULTS

Chemometrics provides a powerful toolbox for bringing quantitative mathematical approaches to complex data such as NMR spectra. Principal Component Analysis (PCA)[2] is *the* fundamental bilinear method in chemometrics. In PCA the data collected on a set of samples is resolved into linear (principal) components, PC's. The first PC becomes the direction in the data which describes most of the variation and the second PC becomes the direction which describes the second most of the variation etc. The PC's are composed of so-called scores and loadings. Loadings contain information about the co-varying variables in the data set and the scores hold information on samples in the data set (amount/concentration of the loadings). The PC's are often plotted against each other in an orthogonal scatter plot giving a 'map' of all the samples in the new coordinate system, the score plot. The major strength of PCA is to provide a quick unsupervised view of the samples and thereby to identify strange samples (outliers) or to discover trends and groups in the samples. Samples where the measurement has failed or in cases where two samples have been switched will manifest as outliers. By decomposing a large data set into a few latent variables it is possible to resolve and quantify important relationships. In this trial we will demonstrate the principles of PCA by application to ^{1}H NMR spectra of a designed set of samples with varying concentration of three chemical compounds from 0 to 100%. If reference data are available it is possible to develop calibration models called partial least squares regression (PLS) models[3]. The bilinear data matrix (X) is resolved in linear components just as in PCA. However, focus is not only to describe the variation between the samples, but also to emphasize variance in data which co-vary (or is correlated) to a reference method which is used for the regression. Thus, PLS regression is a supervised method which may be utilized to develop prediction models that can replace the reference method by a much faster and perhaps a more precise and accurate NMR method.

The experimental design is a tri-axial design of mixtures of the linear alcohols: propanol, butanol and pentanol. Each alcohol component has 21 different levels in steps of 5 from 0 to 100 %[4]. Each sample was prepared from 495 μl of the mixture (Figure 1) and 55 μl of D_2O (with 5.8 mM of TSP-d4 (per-deuterated 3-trimethylsilyl propionate sodium salt).

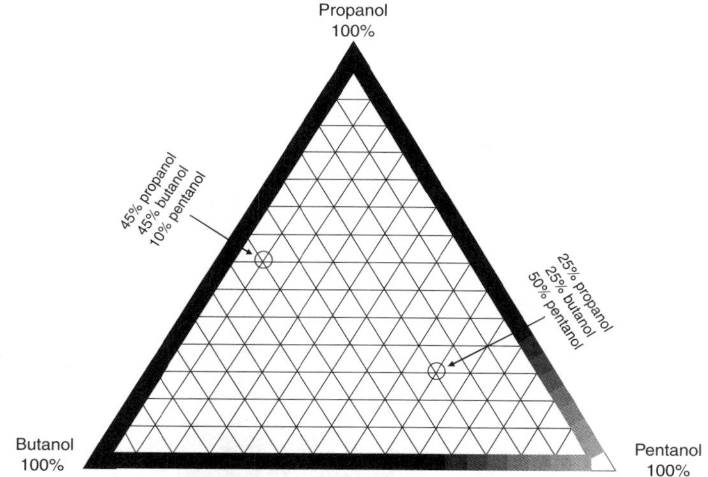

Figure 1 *Tri-axial experimental design of propanol, butanol and pentanol. Each alcohol component has 21 different levels in steps of 5 from 0 to 100 %, 231 samples in total*

^{1}H NMR spectra were recorded for each of the 231 mixtures. The spectra were acquired on a Bruker Avance Ultra Shield 400 spectrometer (Bruker Biospin Gmbh, Rheinstetten, Germany) operating at 400.13 MHz using a broad band inverse probe head equipped with 5 mm (o.d.) NMR sample tubes. Data were accumulated at 298 K using the **zgcppr** pulse sequence[5] with an acquisition time of 4.00 s, a recycle delay of 20 s, 8 scans and a sweep width of 8278.15 Hz, resulting in 64 k complex data points. Pre-saturation was employed throughout the recycle period to obtain sufficient water suppression. In order to secure quantitative measurements the receiver gain was set constant for all the samples. The multivariate data analysis was carried out using the software program LatentiX 1.05 beta (www.latentix.com, Latent5, Copenhagen Denmark).

Figure 2 shows NMR spectra of the 231 alcohol mixtures in the region from 3.5 to about 1 ppm showing four signals. Underneath the full spectrum is an enlargement of the methylene signal coloured in gray according to the propanol content which indicates that increasing propanol content results in decreasing signal intensity. The spectrum of pure propanol (sample 21) yields a triplet at 0.90 ppm from the CH_3, a quintet at 1.55 ppm from CH_2 and triplet at 3.57 from the CH_2 next to the OH group. The corresponding assignments from the pure spectra of propanol, butanol and pentanol are given in Figure 3.

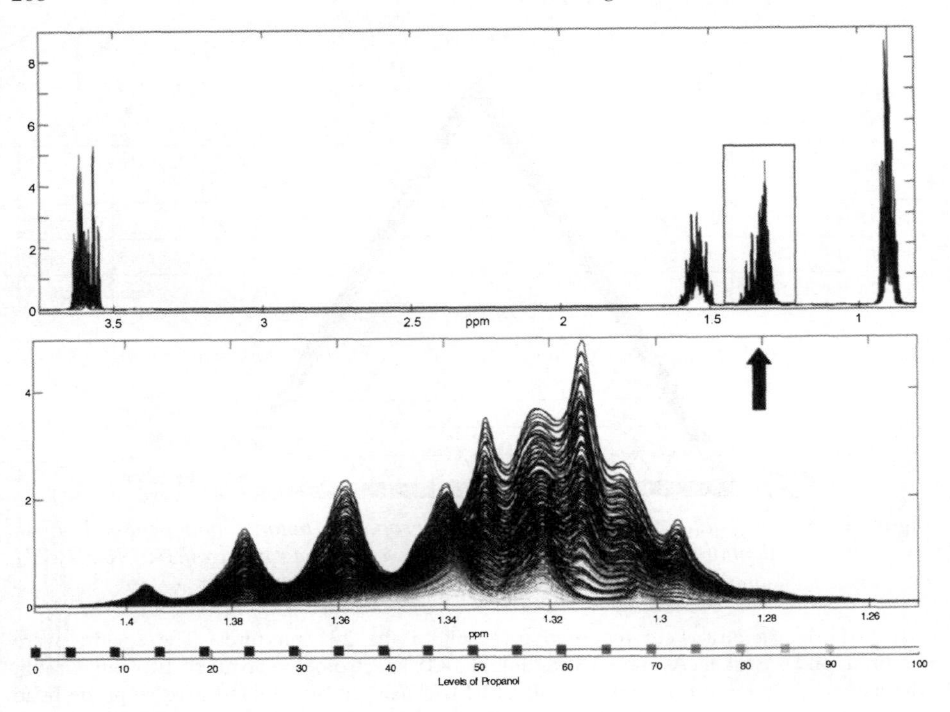

Figure 2 *The ^1H NMR spectra of the samples in the region from 3.5 to 1 ppm (top). Beneath the spectra is an enlargement of the methylene peak coloured by the propanol content*

2.1 PCA results

Using PCA the 231 ^{1}H NMR spectra are decomposed into principal components which describe the systematic variation in the spectra. Figure 4 shows the scores and the loadings of the first two components (PC1 and PC2). The scores are plotted in a scatter plot called a score plot and coloured by the propanol content. PC1 and PC2 describe together 92.5% of the variation in the spectra which in turn tells us that 7.5% of the variation remains to be explained. These two PC1 and PC2 'directions' are displayed in the corresponding loading plot. The ternary experimental design is almost perfectly recovered in the score plot, but with a significant skewness. The reason for this skewness is that the experimental triangle is only fully recovered by adding the third PC, i.e. a third dimension. The third PC explains the remaining 7.5% of the variance in the data (not shown). The loading plot shows the axis 'directions' of the scores. Typically, PC1 describe the primary data structure of the NMR spectra, whereas PC2 describes primarily what makes PC1 different when propanol is added. The fact that the scores and loadings of PC1 are negative is arbitrary, as a negative score multiplied by a negative spectrum yields a positive signal (spectrum).

	$OHCH_2$		CH_2	CH_2CH_3	CH_3
$OHCH_2CH_2CH_3$	3.57 ppm		1.55 ppm	-	0.90 ppm
$OHCH_2CH_2CH_2CH_3$	3.62 ppm		1.53 ppm	1.35 ppm	0.91 ppm
$OHCH_2CH_2(CH_2)_2CH_3$	3.63 ppm		1.55 ppm	1.31 ppm	0.98 ppm

Figure 3 *Assignments of propanol, butanol and pentanol. The chemical shifts of the different protons are given in ppm*

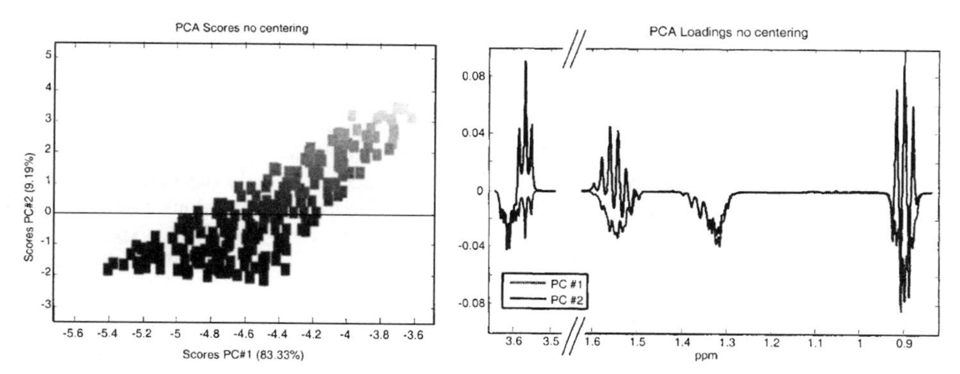

Figure 4 *Scores and loadings plot of the first two principal components from a PCA model calculated on the raw NMR spectra. For increased interpretability the score plot is coloured according to the propanol content. The first two principal components explain 92.5% of the variation*

The most common data transformation prior to PCA is mean centering, i.e. the mean spectrum is subtracted from the individual sample spectra. This simple pre-transformation ensure that the first principal component describes the variation between the samples rather than the direction in which the samples are located. By mean centering the common feature of the samples are removed. In our case mean centering prior to PCA results in a perfect recovery of the ternary experimental design by only the two first PC's, as seen by the score plot in Figure 5.

When the model is mean-centred the first two components describe 98.2% of the total variation in the data. After mean centering apparently only two components are necessary to describe all the variation in the spectra and indeed the remaining 1.8% variation appears non-systematic. The triangular experimental design is now near equilateral, which reflects the direct proportional signal intensities with analyte concentrations. We consider this as a model example of how chemometrics can be used to visualise causal relationships among samples. The fact that the model is now mean-centred is also reflected by the fact that the score plot is centred around origo and the fact that the 'NMR spectral' loadings are now more difficult to interpret and resemble difference spectra rather than real spectra.

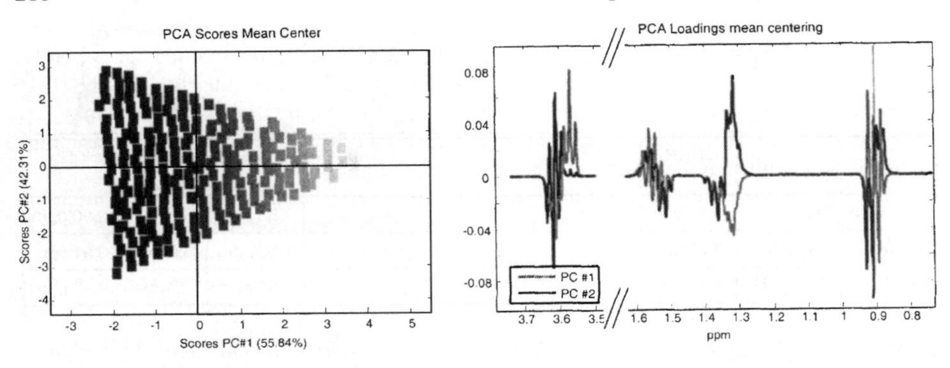

Figure 5 *Scores and loadings plot of the first two principal components from a PCA model calculated on mean-centred NMR spectra. For increased interpretability the score plot is coloured according to the propanol content. The first two principal components explain 98.2% of the variation*

By choosing a specific region of the spectra the PCA model can sometimes be markedly improved. Instead of using the entire NMR spectra with 65.000 variables we will now investigate the variation of a smaller range of the spectra, namely the methylene peak centred at 1.34 ppm and containing only 783 variables. The tremendous strength and robustness of PCA quickly revealed that it was possible to recover the ternary experimental design in the score plot from a model based on only one signal from the NMR spectrum. Figure 6 shows a score and loading plot of a PCA model obtained on the methylene signal alone. The score plot in Figure 6 displays the ternary design reproducing 99.77% of the variance in the methylene data. This may appear as an improvement from the full spectrum model, but in practice this is normal, because a smaller spectral region contains fewer interferences and less spurious noise. The structure of the loadings largely resembles the spectra of butanol (PC1) and pentanol (PC2) which are the only two compounds contributing to the *poly*methylene spectrum (*vide supra*).

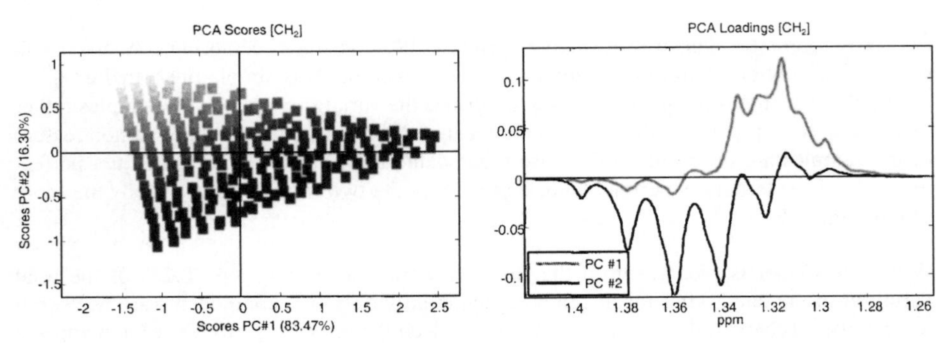

Figure 6 *Scores and loadings plot of the first two principal components from PCA model calculated on the mean-centred methylene NMR signal. For increased interpretability the score plot is coloured according to the propanol content. The first two principal components explain 99.77% of the variation*

Despite the fact, that only two of the three components are physically represented in the data and the model of the methylene peak, the ternary experimental design is fully recovered. This "surprising result" is due to the fact that the intensity of the two components is affected by the presence of the third component by the closure principle, i.e. the concentrations of the three components are added to 100%. The intensity decrease, due to the propanol dilution thereby forms the 'third' direction in the triangle.

If data is affected by non-additive physical effects or other artefacts a transformation of the data may be necessary. Several data pre-transformation methods exist, each specialised in a specific often physical problem. Normalisation is a powerful but strongly manipulating method. There are different normalisation methods; however, it is common that the samples (rows) are summed to a constant value, i.e. the spectra area which demand that the proton density in the samples is identical. In this case, a normalisation of the NMR data clearly deteriorates the reconstruction of the experimental design in the score plot (see Figure 7). Sample No. 21 (located far away from all the other samples) is the 100% propanol sample – with no signal. Even a small amount of butanol or pentanol re-enters the sample to the methylene area. The plot to the right in Figure 7 is the score plot from the PCA model calculated without the 100% propanol sample. The signal intensity is equalised so that the information about the concentration of the third propanol component completely disappears.

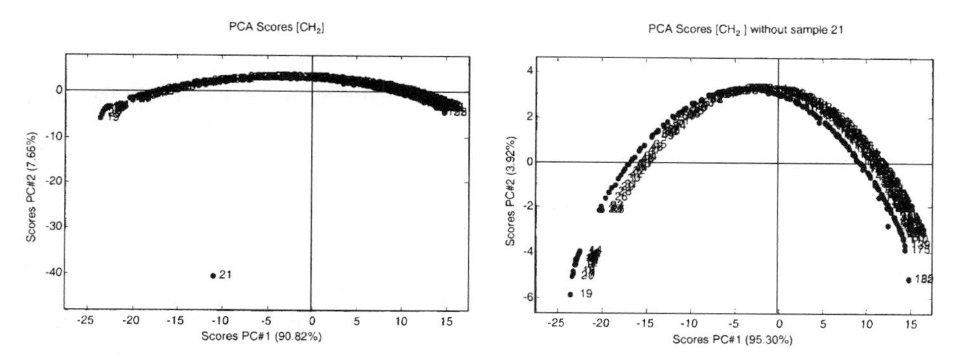

Figure 7 *PCA score plot calculated on the normalised methylene spectra with (left) and without sample 21. Sample 21 is the 100% propanol sample which does not have signal in the examined region.*

2.2 PLS results

PLS regression is the second basic working horse algorithm of chemometrics. It works quite similarly to PCA, but its scope is to regress (or force) the result in a given direction (reference method) and is thus called a supervised method. In this paper we aim to demonstrate PLS by using the known alcohol concentrations (from the experimental design) to develop PLS regression models. The number of PC's in the PLS model is determined using systematic cross-validation with five segments, i.e. the sample set is divided in to five which in turn gives five times 46 samples that are excluded "at a time" before re-entering into the model in order to estimate the prediction error. The calibration models calculated on this type of experimental design data usually provide excellent models and indeed the regression models from the methylene signal to the alcohol concentrations performed very well with correlation coefficients close to 1 and new samples could be predicted with a precision of only about 1.3 % for each alcohol using only 2 PLS components (optimal number of PC's determined by cross-validation).

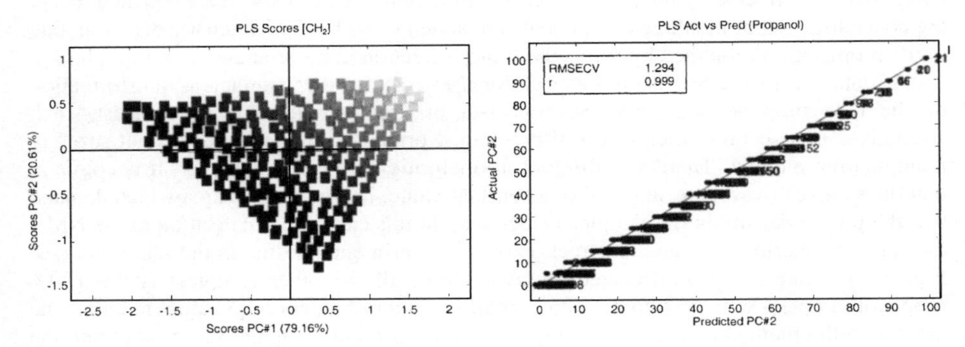

Figure 8 *PLS score plot of methylene signal (left). For increased interpretability the score plot is coloured according to the propanol content. The first two PCs represent 99.77% of the variation. Predicted versus actual/measured plot of the PLS model on propanol yielded a prediction error of only 1.3% and a correlation coefficient of 0.999 (right)*

It is interesting to note that the PLS score plot in Figure 8 represents exactly the same variation in the X-data as the PCA score plot did in Figure 6, namely 99.77%; but in this case the (PLS) score plot is a little more equilateral. It is furthermore of interest that the excellent regression model is made to a component (propanol) which does not contribute to the spectra in the investigated region – making the model an indirect model relying on closure and inverse correlation. However, this is a normal situation when analysing complex foods, metabonomics and nutrigenomics data. When examining the loadings in Figure 9 we observe that the PLS loadings to a lesser extent than the PCA loadings in Figure 6 resemble the pure spectrum of butanol and pentanol in the methylene region. However, the loadings are not identical to the spectra of butanol and pentanol and this is due to the mean centering of the data and to the orthogonalisation restriction in the PCA and PLS algorithms.

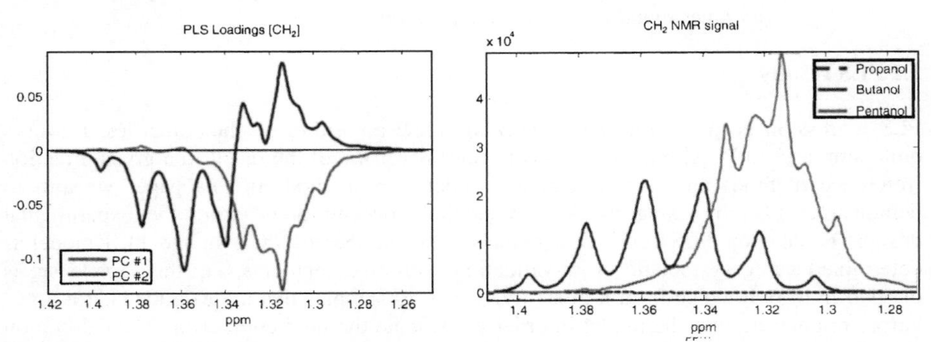

Figure 9 *PLS loading plot of the first two components from the PLS model (left) and pure NMR spectra of butanol and pentanol (right). Propanol is not giving any signal in this region*

3 CONCLUSIONS

By applying the basic chemometric algorithms PCA and PLS to a well defined ternary experimental design of ^{1}H NMR spectra we hope to have demonstrated the potential and characteristics of chemometric multivariate data analysis. The perhaps greatest advantage of chemometrics is the simplicity by which even large data structures are analysed and its exploratory nature. Although the link to food analysis by NMR is rather vague we hope to have demonstrated the potential of chemometrics to investigate and explore static as well as dynamic food systems using NMR spectroscopy as the sensor principle. The combination of NMR and chemometrics has tremendous potential in the nutrigenomics and metabonomics fields for exploring patterns of biomarkers for diseases and food intake in biofluids, but a future in process analytical technology (PAT) in the food and medico industries is also highly probable.

4 REFERENCES

1. H. Serrai, L. Nadal, G. Leray, B. Leroy, B. Delplanque, and J.D. de Certaines. *NMR in Biomed.*, 1998, **11,** 273

2. H. Hotelling. *J. Educ. Psychol.*, 1933, **24,** 417

3. S. Wold, H. Martens, and H. Wold. *Lecture Notes in Mathematics*, 1983, **973,** 286

4. T. Næs and T. Isaksson. *NIR News*, 1992, **3,** 7

5. A. Bax. *J. Magn. Reson.*, 1985, **65,** 142

PHYTIC ACID DEGRADATION BY PHYTASE AS VIEWED BY ^{31}P NMR AND MULTIVARIATE CURVE RESOLUTION

M.M. Nielsen, N. Viereck and S.B. Engelsen

Quality & Technology, Department of Food Science, The Royal Veterinary and Agricultural University, Rolighedsvej 30, DK-1958 Frederiksberg C, Denmark

1 INTRODUCTION

There is a growing interest amongst consumers and industry to ensure that our food is safe, healthy and nutritious. To achieve high standards this will require rapid, sensitive and reliable analytical methods. Spectroscopy is very well suited for this purpose and different types of spectroscopy have already been utilized in industry, such as fluorescense, infrared and near-infrared spectroscopy, including low-field NMR relaxometry. So far there has not been great tradition to utilize high-resolution (HR) NMR spectroscopy within food science and food industry. The fact that NMR spectroscopy is non-destructive and provides quantitative and structural information as well as information on water mobility and distribution about the entire sample volume makes this method particularly useful for quality control in the food industry.

In order to enhance interpretation and the quantitative information for measurement of complex biological/food systems it is often necessary to combine the spectroscopic results with advanced multivariate (chemometric) methods which open up for a great unutilized potential for spectroscopic quality control of food by exploring food systems using unsupervised data technologies which require no *a priori* information. Chemometrics is also referred to as "statistics without tears", because it facilitates handling of large data sets and deals efficiently with real-world multivariate data, taking advantage of the previously feared co-linearity of spectral data and for providing the possibility of projecting the data into a few dimensions via a graphical representation[1]. Multivariate data visualization of real-life data sets is perhaps the single most important feature of chemometrics. It is important to underline that there are NO univariate real-world problems – this has been most definitively stated by the traditionally conservative FDA in its PAT (Process Analytical Technology) guidance for the Pharmaceutical Industry: "*Traditional one-factor-at-a-time experiments do not effectively address interactions between products and process variables*".

With direct reference to metabonomics[2] we have introduced the term *bromatonomics* which covers "quantitative measurements of functional quality parameters of food as studied by NMR on intact food systems and evaluated by multivariate data analysis". In this paper we will exemplify the application of simple unsupervised chemometric methods to ^{31}P NMR spectra in order to study the dynamics of phytic acid being degraded by phytase – an important reaction for the mineral availability in food and feed. Phytic acid (myo-inositol 1,2,3,4,5,6-hexakis dihydrogen phosphate, IP6), which is shown in Figure 1, forms complexes with different divalent and trivalent cations as well as with proteins and starch, and the solubility of the complexes is pH-dependent[3]. Most phytic acid-mineral complexes, called phytate, are soluble at a pH less than 3.5 with maximum insolubility occurring between pH 4 and 7[4, 5]. The complexes depend on the numbers of phosphate groups on the inositol ring. The fewer phosphate groups on the ring, the weaker the complexes become. The insolubility of the phytate complexes is probably the major reason why phytic acid historically is considered an anti-nutrient because it binds the essential dietary minerals and thus decreases their bioavailability[6]. Phosphorus in the form of phytate is generally not bioavailable to non-ruminant animals, because they lack the digestive enzyme phytase which is required to separate phosphorus from the phytate molecule. The approximate pH of the intestine, where absorption of metal ions occurs, coincides with the pH at which these complexes precipitate.

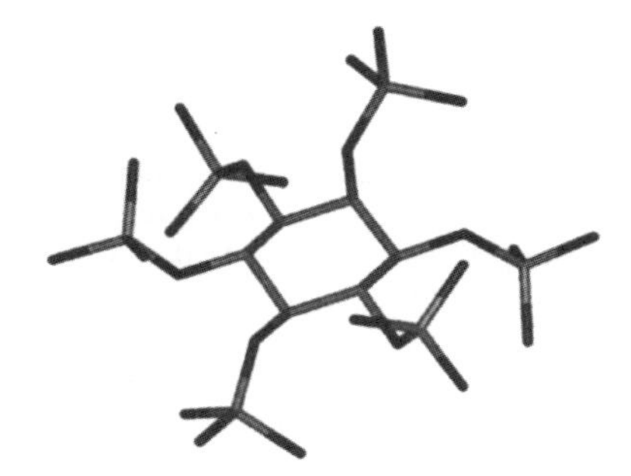

Figure 1 *The heavy atom structure of phytic acid. Only the phosphate group on C-2 is axial.*

Phytase is the enzyme that catalyses the stepwise hydrolysis of phytic acid to lower inositol phosphates (myo-inositol pentaphosphate, IP5 to myo-inositol mono phosphate, IP1, Figure 2).

IP6 → IP5 → IP4 → IP3 → IP2 → IP1 → *myo*-inositol

Figure 2 *Stepwise dephosphorylation of IP6 (phytate)*

There are two international classified phytases: 3-phytase (EC 3.1.3.8) and 6-phytase (EC 3.1.3.26). The enzymes are named after the position of the first phosphorester bond of the phytate to be hydrolyzed. 3-phytase seems to be of microbial origin, while 6-phytase is synthesized by plants[7]. A 4-phytase seems to be present in cereals as well[8, 9]. This enzymatic activity produces available phosphates that are considered important in upgrading the nutritional quality of phytate-rich foods and feeds[10]. In modern agriculture, non-ruminant livestock such as swine and poultry are fed mainly grains like soybeans and maize. Because phytate from these grains is unavailable for absorption, the unabsorbed phytate passes through the gastrointestinal tract, elevating the amount of phosphorus in the manure. Excess phosphorus excretion can lead to environmental phosphorus pollution.

2 METHODS

2.1 Samples

The ^{31}P NMR model experiment samples were prepared by dissolving a commercial phytase enzyme isolated from wheat (Biofeed, Novozymes) with a sodium phytate solution (7.5mg/ml, 30mg/ml EDTA) adjusted to pH 5.3 with sodium acetate buffer (0.25M). The enzyme had an activity of 0.01 U/ml in the solution (1 unit of enzyme activity liberates 1 μmol of inorganic phosphorus from a 1.5 mM phytate solution per min. at pH 5.15 at 55°C).

2.2 ^{31}P NMR

^{31}P NMR spectra were recorded to study the phytase reaction every 9 min. over a period of 14 hours. The spectra were acquired on a Bruker Avance Ultra Shield 400 spectrometer (Bruker Biospin Gmbh, Rheinstetten, Germany) operating at 161.98 MHz using a broad band inverse probe head equipped with 5 mm (o.d.) NMR sample tubes. Data were accumulated at 298.5 K using the zgpg30 pulse sequence with an acquisition time of 10.1 s., a recycle delay of 10 s, 32 scans and a sweep width of 3238.34 Hz, resulting in 64 k complex data points. In order to secure quantitative measurements the receiver gain was set constant for all consecutive NMR measurements.

2.3 Unsupervised exploratory analysis

One of the main advantages of applying multivariate chemometric data analysis to collinear spectroscopic data is the possibility of carrying out an exploratory inductive investigation[11]. The universal basic chemometric algorithm principal component analysis (PCA)[12] is a most useful tool for this purpose. PCA is based on the calculation of underlying latent data structures using a two-dimensional data strategy, i.e. measuring a series of samples and finding common latent data structures and individual scores or concentrations. Common to such bilinear models is that an entire matrix, with each row being the measurement from one sample, must be acquired. PCA finds the main variation in a multidimensional data set by creating new linear combinations (orthogonal) of the raw data that approximate the original data set in a least squares sense, a method which is especially well suited to highly collinear data, as is the case in most spectroscopic or instrumental techniques. The model to be solved is $X = T*P^t$ in which the data matrix (samples × spectra) $\mathbf{X}$ is decomposed into a lower dimensional score matrix ($\mathbf{T}$) and a loading matrix ($\mathbf{P}$). In this way the information in $\mathbf{X}$ is projected onto a lower dimensional subspace where the loading vectors for the principal components (PC) can be regarded as pure mathematical spectra that are common to all the measured spectra. What makes the individual raw spectra different are the amounts (scores) of hidden spectra (loadings). PCA can be considered as the first amendment in exploratory data analysis due to its extraordinarily robust data reduction, its low level of *a priori* assumptions and its data presentation capabilities.

2.4 Constrained Alternating RegreSsion (CARS)

Alternating regression, also called multivariate curve resolution (MCR)[13], is a mathematical method for curve resolution that does not impose the orthogonality constraint as utilized in the PCA algorithm. Analogous to PCA, the input to the algorithm is a matrix X with spectral measurements of the relevant mixtures to be resolved, and the output is two matrices: C that contains the estimated concentrations of the mixture components, and S that contains the pure spectra of the mixture components. In this study CARS is implemented with non-negative constraints[14] on both the concentrations and the spectra. The model to be solved is $X = C*S^t$, where t is transposed and dimensions are: X (n samples, m wavelengths), C (n samples, p pure components), S (m wavelengths, p pure components).

The pseudo code for the implemented method is as follows:

```
While changes in S observed
  for i = 1 to number of samples
    Solve C = X/St, with non-negativity constraints
  end
  for i = 1 to number of pure components
    Solve S = X/C, with non-negativity constraints
  end
end
```

In contrast to PCA, CARS require a guess of S in order to initiate the algorithm, but with the presented data the algorithm proved rather reproducible regardless of the initial guess.

The basic chemometric data analysis and visualization of spectra was made using the data workbench and PCA algorithm from the public domain. chemometric software LatentiX version 1.05 (www.latentix.com). The CARS algorithm (www.kvl.models.dk) was implemented under MatLab Version 13 (MathWorks Inc., Natic, USA).

3 RESULTS AND DISCUSSION

The degradation of phytic acid in a model system in which phytase was added to an aqueous solution was investigated by a time series of ^{31}P NMR spectra. Figure 3 shows the NMR spectra as a function of time. In the beginning before the phytase action (very light grey) four ^{31}P peaks are present (0.14, 0.47, 1.06 and 1.76 ppm) with the intensity ratios 1:2:2:1 resulting from the six P atoms in the phytic acid molecule (IP6) in which 2 pairs of phosphate atoms are degenerated due to the symmetry[15]. However, after addition of phytase the pattern and intensity of the peaks change as a consequence of degradation of IP6 to lower inositol phosphates. First, five new peaks are observed in the spectra which may be due to the formation of one specific IP5, because only five equally strong peaks are formed. This is because it is an asymmetric molecule which is formed where carbons 2 and 5 still are phosphorylated and in good agreement with the fact that we have been using a wheat phytase that is specific for cleavage of carbon 6. The enzymatic process gets more and more complex, but in agreement with earlier studies, one dominating isomer is formed for the penta-, tetra-, tri-, di- and monophosphates, which is outlined more clearly in Figure

6[15]. The phytase process continues for 14 hours and ends with four peaks (0.38, 0.42, 2.45 and 2.77 ppm) shown by the black line in the spectra. The four peaks remaining at the end corresponds to the three isomers of IP1 that remain after the hydrolysis and to inorganic phosphate (0.38 ppm).

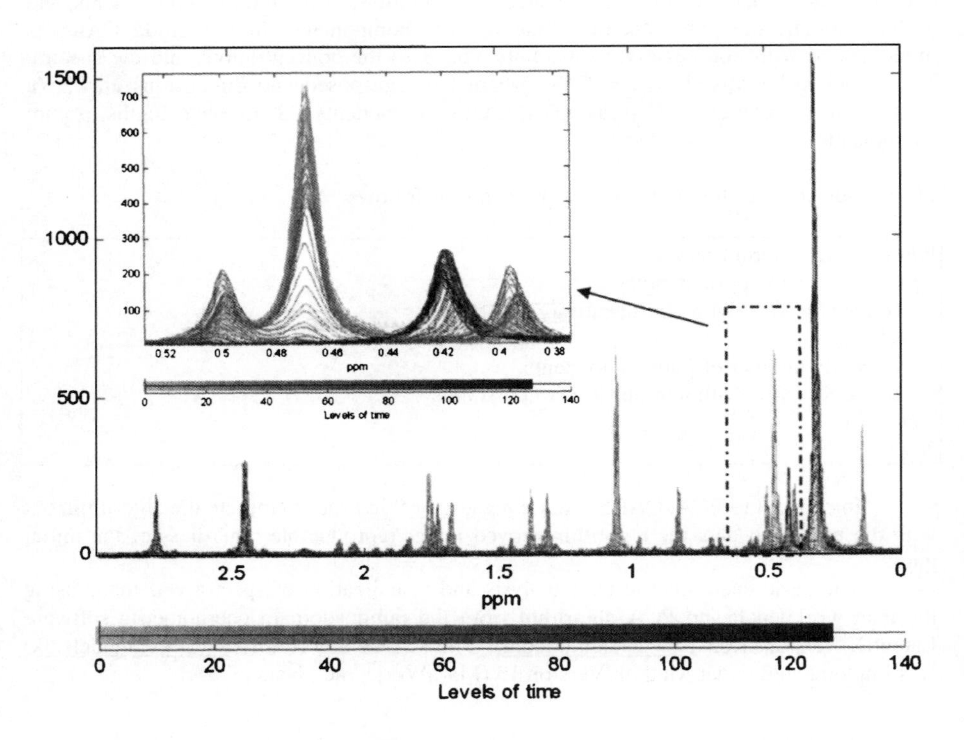

Figure 3 *31P NMR spectra showing the hydrolysis of pure sodium phytate with phytase for 19 hours.*

To illustrate the changes in the ^{31}P spectra over time the area from 0.37 to 0.57 ppm is zoomed. From the figure it is apparent that three peaks decrease (including IP6 at 0.47 ppm) while one increase. This pattern of changes continues as IP5 degrades to IP4 and IP4 to IP3 and so on until only IP1 and inorganic phosphorus remain.

Figure 4 displays the results from a PCA of the time series ^{31}P spectra. The score plot, which sums up 94% of the total variation in the spectral data, has a typical horseshoe shape common to (mean-centered) sample reactions. This simplicity is not anticipated for the IP6 to IP1 complex reactions, which is already reflected if we add a third PC in a 3D score plot (see Figure 5). The additional 3% variation described by PC3 gives a complex loop pattern of the scores, indicating a rather complex score pattern.

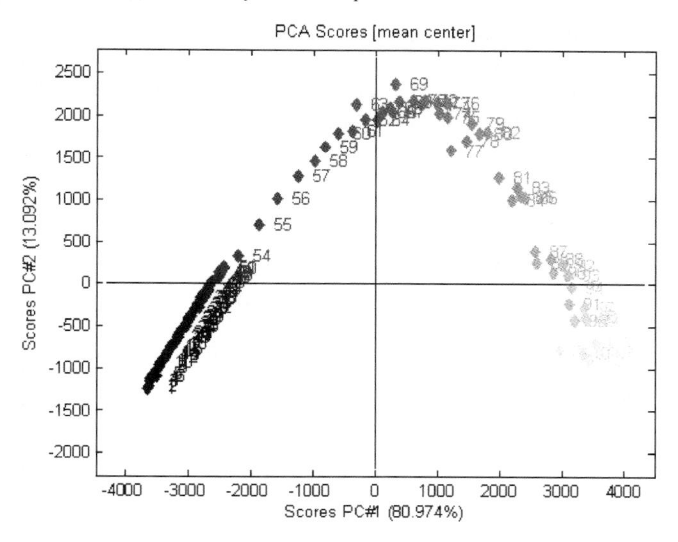

Figure 4 *PCA score plot of the mean-centered ^{31}P spectra. PC1 and PC2 together explain 94% of the variation in the NMR spectra.*

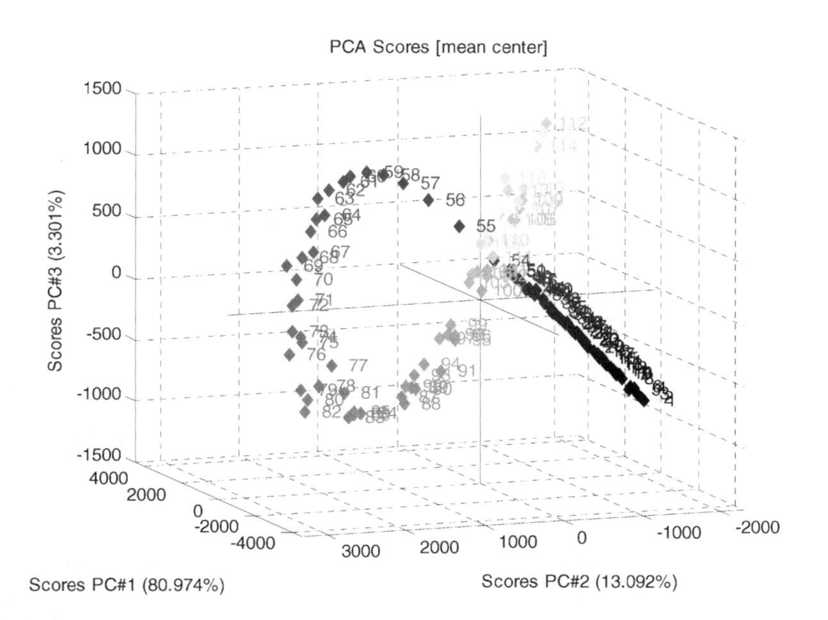

Figure 5 *3D PCA score plot of the mean-centered ^{31}P spectra. PC1, PC2 and PC3 together explain 98% of the variation in the NMR spectra.*

Instead of modeling the 31P NMR spectra by PCA a more realistic model may result from the CARS or multivariate curve resolution method which does not impose the orthogonality constraint. Figure 6 displays the result of a six component CARS analysis of the ^{31}P NMR time series of the phytic acid degradation.

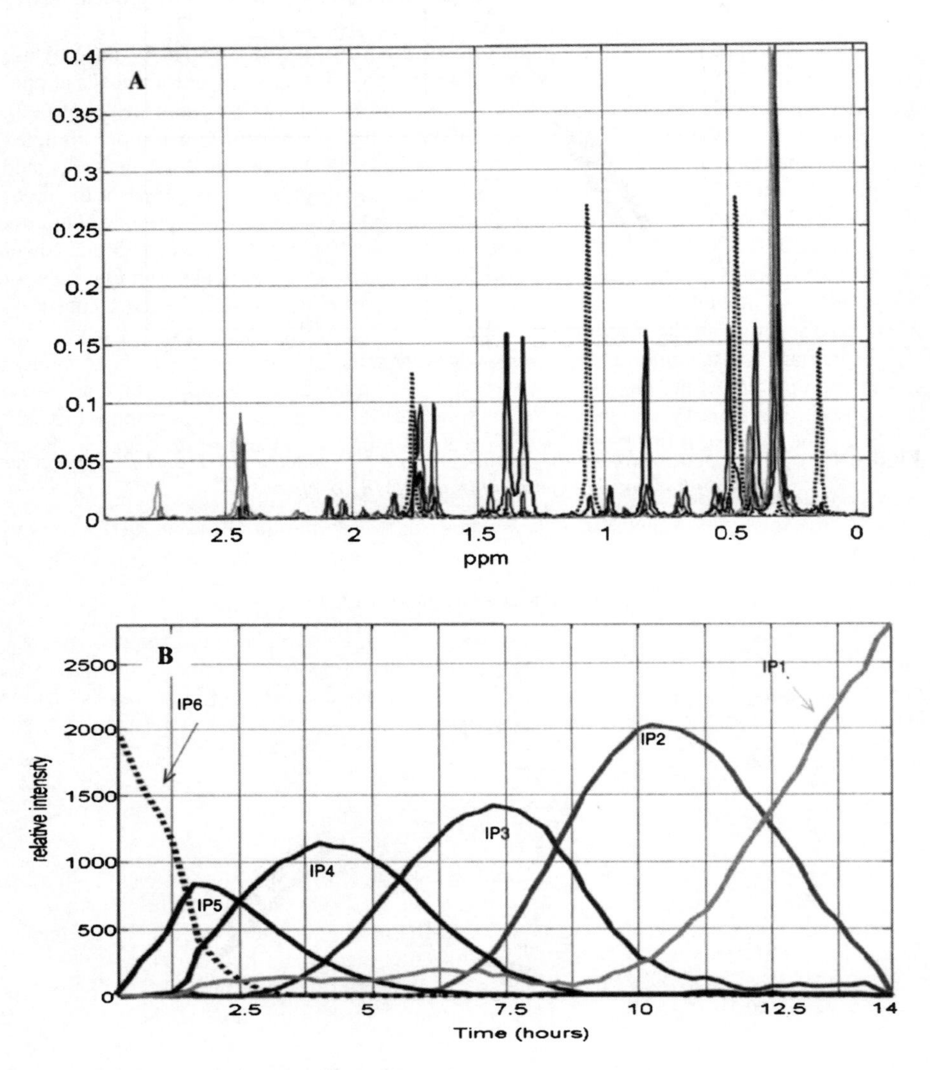

Figure 6 *(A)Pure ^{31}P NMR spectra from the CARS model. Dotted line=IP6, Black line=IP5, Dark gray=IP4, Gray=IP3, Light gray=IP2 and Very light gray=IP1. (B) Plot of the relative intensity as a function of time. The same colors are used as identification for the different inositol phosphates as in (A)*

In the case of these NMR data the CARS method proved very reproducible independent of the initial guesses (random or spectra) of S. The model using six components gave the best description of the time series NMR spectra. A five-component model described less of the total variance and the seven-component model began to divide obvious single components. Figure 6 shows the result of the optimal six-component CARS model.

The six spectra of the compounds resolved from the time series of the phytase reaction using CARS is displayed in Figure 6A. The dotted line showing four peaks at ppm 0.14, 0.47, 1.06 and 1.76 represents IP6. IP5 is represented by five main peaks at 0.30, 0.38, 0.48, 0.83, 1.32 and 1.37 ppm. From then on the reaction becomes more complex because of the many possible configurations of phytate. In the case of IP4 3 peaks located at 1.68, 1.73 and 1.75 ppm dominate, but also many small peaks also coexist in the areas around 1.4, 2.1 and 2.4 ppm. These patterns are in good agreement with the rare literature found about 31P NMR on pure phytate model systems[15-17]. In the case of IP3 and lower inositol phosphates it becomes very difficult to separate the peaks without use of chemometrics; however, the multivariate curve resolution program CARS makes it possible to separate all the components.

Perhaps most important is the result in Figure 6b which shows the complete degradation pattern of the phytase reaction using real multivariate experimental data. From the figure it is evident that already after 2.5 hours all IP6 is degraded. After approximately 1.5 hours the formation of IP5 (black line) reaches a maximum and for the next six hours undergoes degradation to IP4 which reaches a maximum after four hours and is fully degraded after ten hours. This is in agreement with Frølich and coworkers[15] who did a similar experiment and found that IP4 was degraded after approx. 10 hours. According to the literature, phytic acid should be degraded to at least IP3 before the complex bound minerals are released and thus become bioavailable[18, 19]. This implies that it takes approximately 10 hours (at the given concentration and pH) to degrade phytate to a form where the mineral complexes are no longer so strong and the minerals thus bioavailable for human and animals. IP3 has a lifetime of approximately 10 hours; for IP2 it is approximately 8 hours. IP1 is formed during the entire hydrolysis, but increases most after 10 hours.

4 CONCLUSIONS

The ^{31}P NMR method is a most direct and useful method to describe the degradation of phytic acid to lower inositol phosphates by the action of the enzyme phytase. The use of chemometric and CARS visualizes and helps in the interpretation of the results. By means of LatentiX it has been possible to visualize the time-dependent hydrolysis of phytic acid and by PCA the complexity of the phytic acid is shown in the score plots. By modeling the spectra in CARS it is possible to identify and quantify each of the inositol phosphates.

References

1. E. Micklander, L.G. Thygesen, H.T. Pedersen, F. van den Berg, R. Bro, and S.B. Engelsen. Multivariate analysis of time domain NMR signals in relation to food quality. *In* Magnetic Resonance in Food Science: Latest Developments. *Edited by* G.A. Webb, P.S. Belton, and D.N. Rutledge. 2003. p. 239.
2. J.C. Lindon, J.K. Nicholson, E. Holmes, and J.R. Everett. *Concepts Magn. Reson.*, 2000, **12,** 289
3. R. Lásztity and L. Lásztity. Phytic Acid in Cereal Technology. *In* Advances in Cereal Science and Technology. *Edited by* Y. Pomeranz. St. Paul, Minnesota, USA. 1990. p. 309.
4. R. Siener, H. Heynck, and A. Hesse. *J. Agric. Food Chem.*, 2001, **49,** 4397
5. N.M. Tamim and R. Angel. *J. Agric. Food Chem.*, 2003, **51,** 4687
6. Q.C. Chen and B.W. Li. *J. Chrom. A*, 2003, **1018,** 41
7. C. Centeno, A. Viveros, A. Brenes, R. Canales, A. Lozano, and C. de la Cuadra. *J. Agric. Food Chem.*, 2001, **49,** 3208
8. R. Greiner, M.L. Alminger, and N.G. Carlsson. *J. Agric. Food Chem.*, 2001, **49,** 2228
9. B.Q. Phillippy. Stability of plant and microbial phytases. *In* Food Phytases. *Edited by* M.B. Reddy and S.K. Sathe. New York. 2002. p. 107.
10. M. De Angelis, G. Gallo, M.R. Corbo, P.L.H. McSweeney, M. Faccia, M. Giovine, and M. Gobbetti. *Int. J. Food Microbiol.*, 2003, **87,** 259
11. L. Munck, L. Nørgaard, S.B. Engelsen, R. Bro, and C.A. Andersson. *Chemom. Intell. Lab. Syst.*, 1998, **44,** 31
12. H. Hotelling. *J. Educ. Psychol.*, 1933, **24,** 417
13. R. Tauler. *Chemom. Intell. Lab. Syst.*, 1995, **30,** 133
14. R. Bro and S. DeJong. *J. Chemo.*, 1997, **11,** 393
15. W. Frølich, T. Drakenberg, and N.G. Asp. *J. Cereal Sci.*, 1986, **4,** 325
16. W. Frølich, N.M. Wahlgren, and T. Drakenberg. *J. Cereal Sci.*, 1988, **8,** 47
17. P.A. Kemme, A. Lommen, L.H. De Jonge, J.D. Van der Klis, A.W. Jongbloed, Z. Mroz, and A.C. Beynen. *J. Agric. Food Chem.*, 1999, **47,** 5116
18. M. Brune, L. Rossander-Hultén, L. Hallberg, A. Gleerup, and A.S. Sandberg. *J. Nutr.*, 1992, **122,** 442
19. A.S. Sandberg, M. Brune, N.G. Carlsson, L. Hallberg, E. Skoglund, and L. Rossander-Hultén. *Am. J. Clin. Nutr.*, 1999, **70,** 240

Subject Index